SUSTAINABLE
LANDMARKS

可持续性地标建筑（中）

石大伟 主编

中国林业出版社

ARCHITECTURES

建筑设计

STOP LINE

recreational complex
Curno, Italy, 1993-1995

地点　Curno，贝加莫，意大利
项目　娱乐综合建筑物
客户　Golf Parco dei Colli
结构　Studio Myallonnier
系统　Studio Armondi
规划　1993年
建设　1995年
成本　11,000,000欧元
建筑面积　10,000平方米
体量　60,000立方米
承包商　Falgari Marmi & C. S.r.l.-F.lli Beltramini

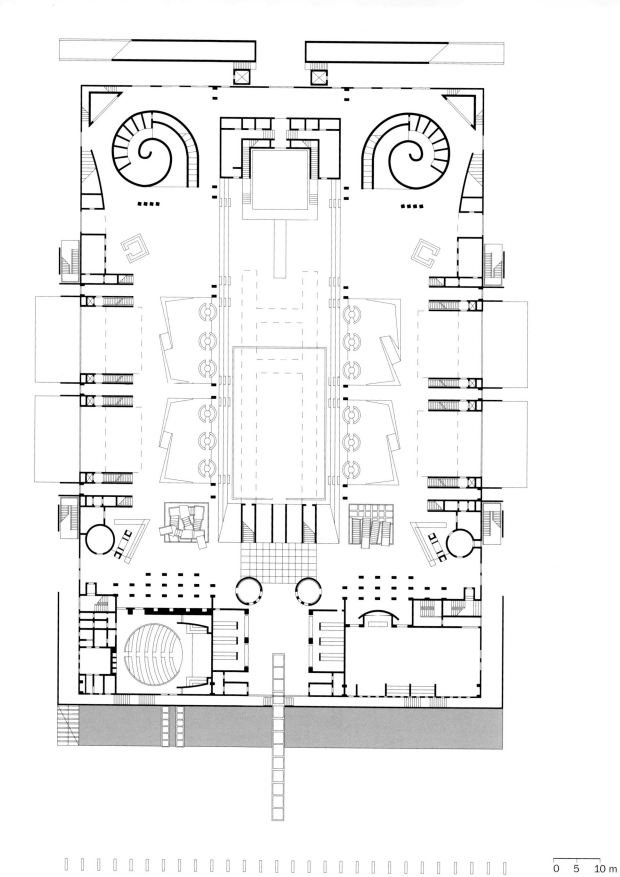

ground floor plan
一层平面

0 5 10 m

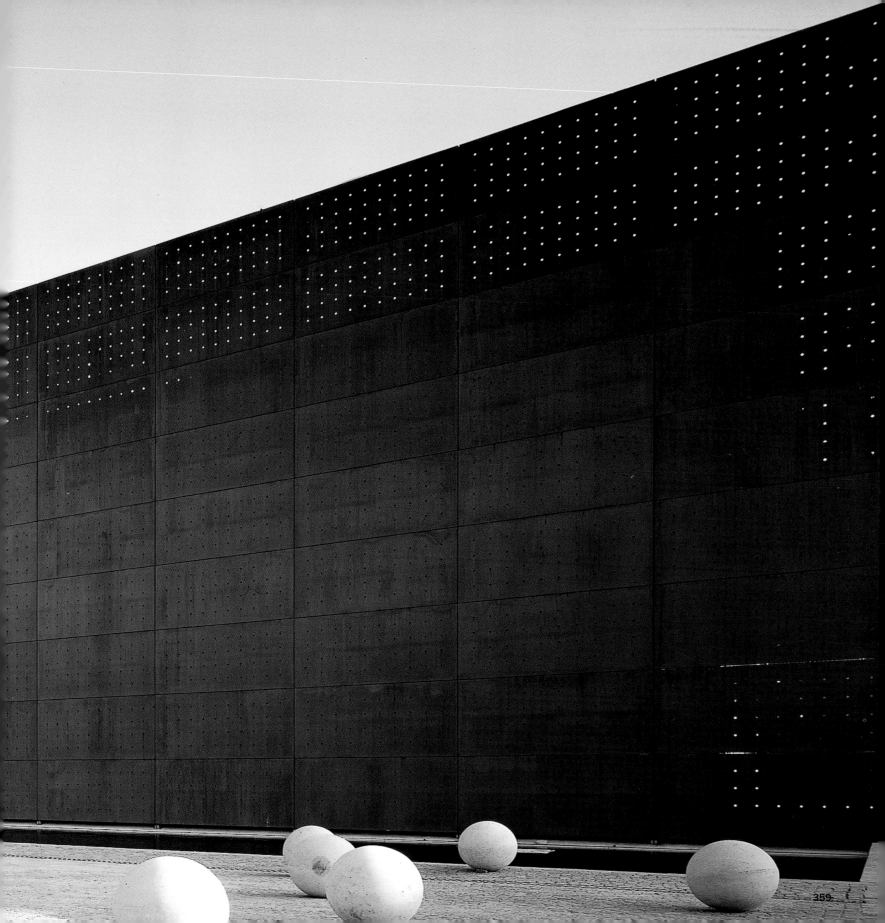

POOL IN MONTE BUE

Cene, Bergamo, Italy, 1995-1996

地点   Cene，贝加莫，意大利
项目   游泳池
客户   私人客户
规划   1995年
建设   1996年
成本   300,000欧元
承包商   Madaschi

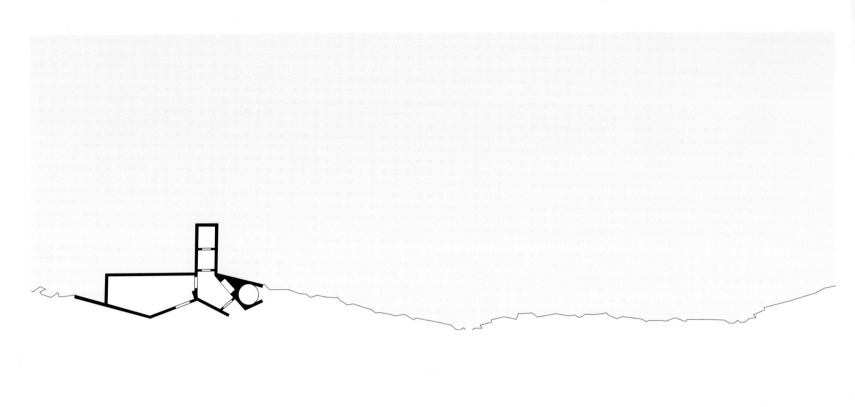

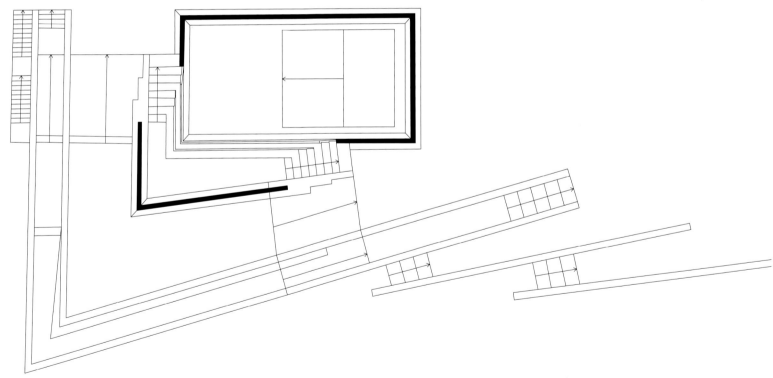

ground floor plan
一层平面

0   5   10 m

SQUARE AND PUBLIC FACILITIES

Merate, Lecco, Italy, 1996-2011

地点	Merate，莱科，意大利
项目	公共广场和公共设施
客户	Merate市政府
结构	Studio Myallonnier
系统	Studio Armondi
规划	1996年
建设	在建 1999年-2011年
成本	3,200,000欧元
占地面积	11,260平方米
建筑面积	5,280平方米
承包商	CIAS Group - SGC Italia S.p.A. - Giacomo Zenga Costruzioni S.a.s

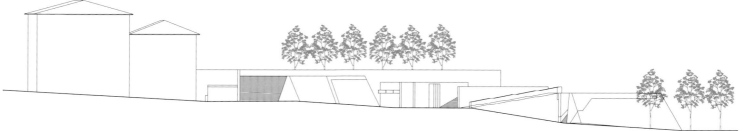

section A-A'
A-A'剖面

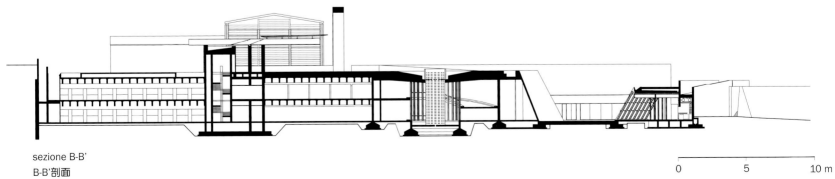

sezione B-B'
B-B'剖面

0　　　　5　　　　10 m

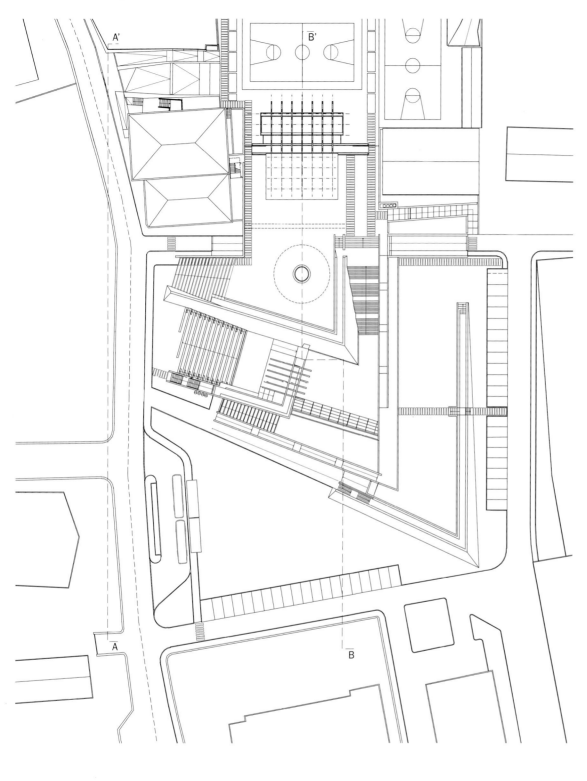

A'

B'

A

B

site plan
位置图

0   5   10 m

389

CURNO LIBRARY AND AUDITORIUM

Curno, Bergamo, Italy, 1996-2009

地点	Curno，贝加莫，意大利
项目	图书馆和礼堂
客户	Curno市政府
结构	Studio Myallonnier
系统	Studio Armondi
规划	1996年
建设	1999年-2009年
成本	2,000,000欧元
建筑面积	1,960平方米
体量	8,200立方米
承包商	Viola Costruzioni

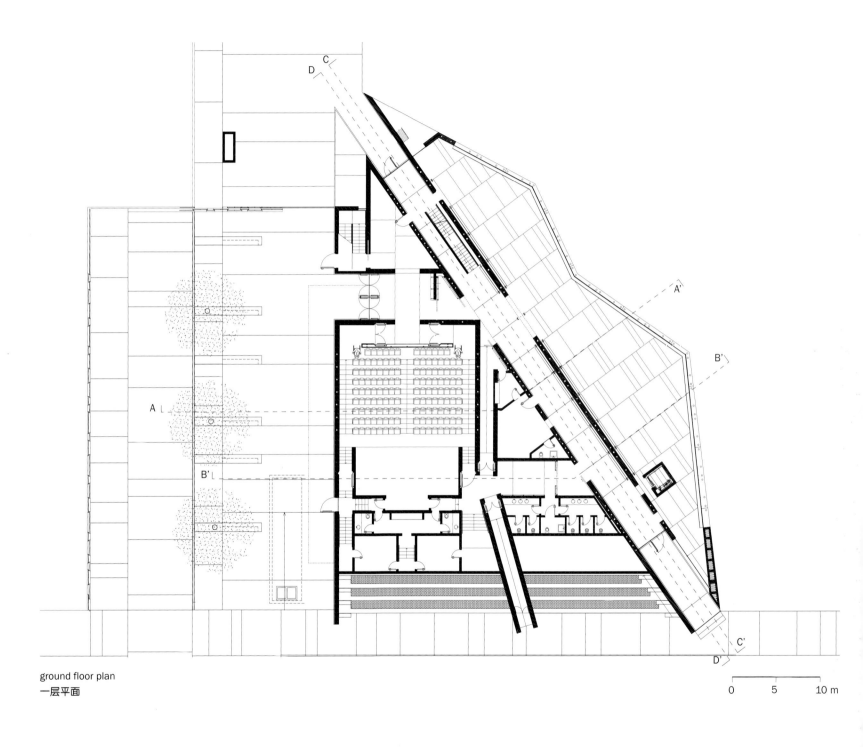

ground floor plan
一层平面

0   5   10 m

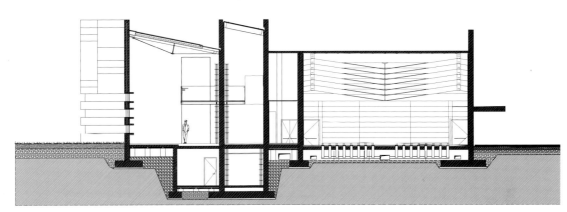

section A-A'
A-A'剖面

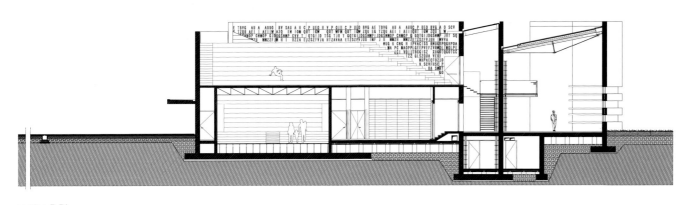

section B-B'
B-B'剖面

0   5   10 m

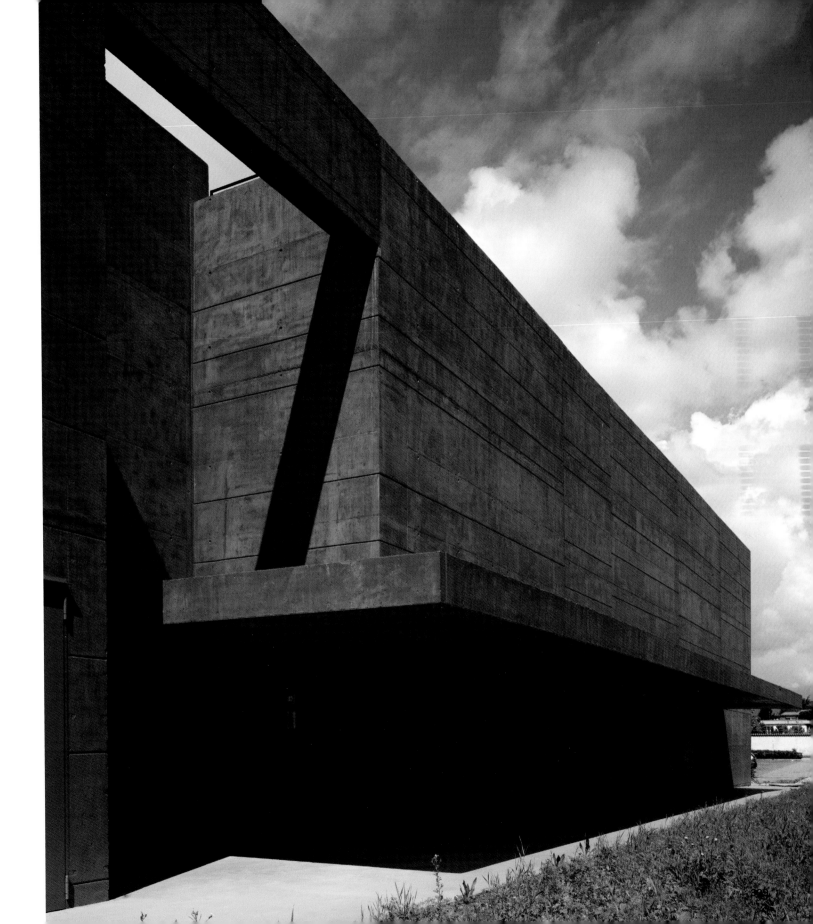

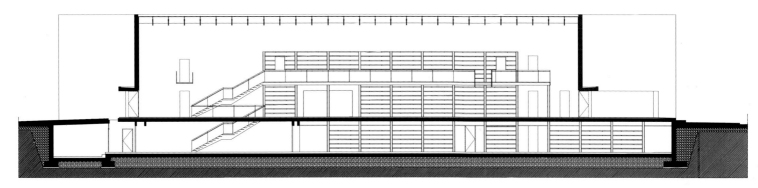

section C-C'
C-C'剖面

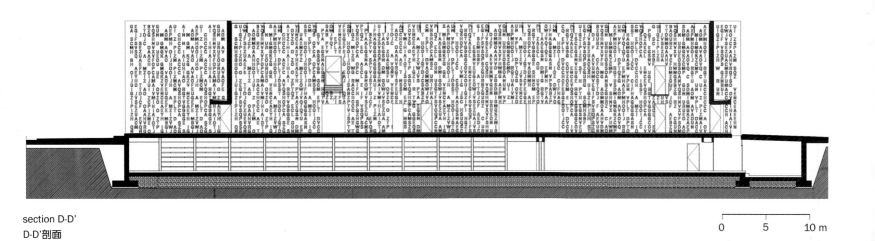

section D-D'
D-D'剖面

0   5   10 m

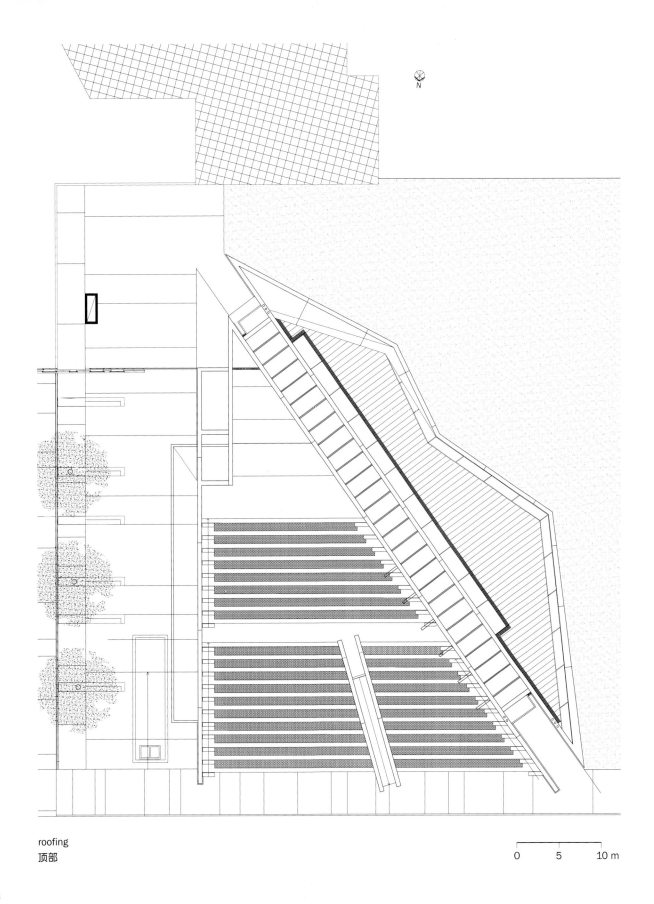

roofing
顶部

0   5   10 m

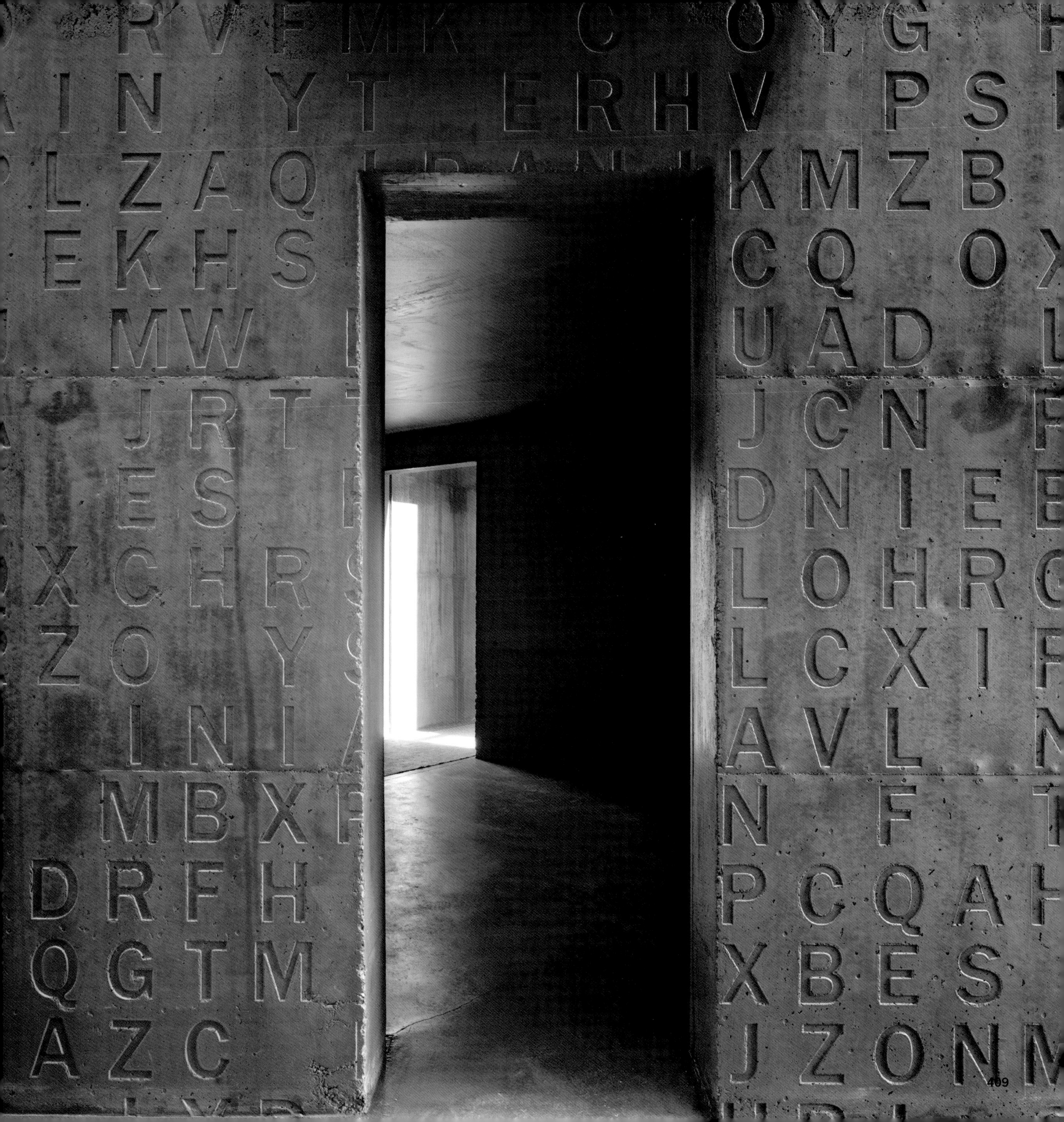

LEFFE HOUSE
Leffe, Bergamo, Italy, 1997-1999

地点	Leffe，贝加莫，意大利
项目	独户房屋
客户	私人客户
结构	Gianfranco Calderoni
规划	1997年
建设	1999年
成本	600,000欧元
占地面积	90平方米
建筑面积	280平方米
体量	800立方米
承包商	Madaschi

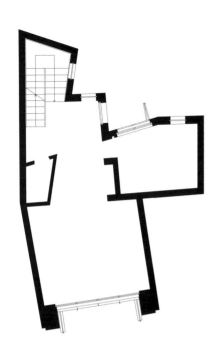

ground floor plan
一层平面

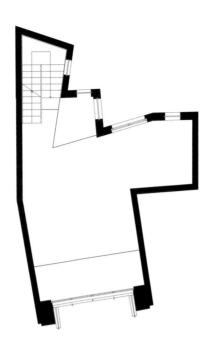

first floor plan
二层平面

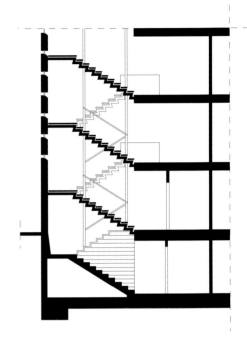

cross section
横剖面

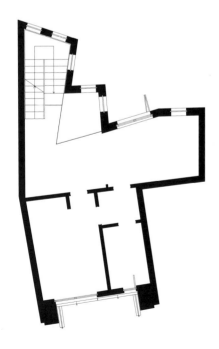

second floor plan
三层平面

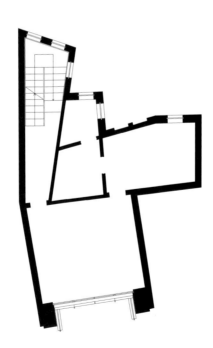

third floor plan
四层平面

0 1   5 m

419

RESIDENTIAL AND COMMERCIAL COMPLEX

Tavarnuzze, Florence, Italy, 2000

地点　Tavarnuzze，佛罗伦萨，意大利
项目　住宅和商店
客户　Bruno Cecchi S.p.A.
结构　Vega Ingegneria
机械系统　Studio Mancini
电气系统　Xenia s.r.l. Alessandro Lepri
规划　2000年
建设　在建
成本　8,000,000欧元
占地面积　6,600平方米
建筑面积　5,256平方米
体量　17,500立方米

longitudinal section
纵剖面

site plan
位置图

0   5   10 m

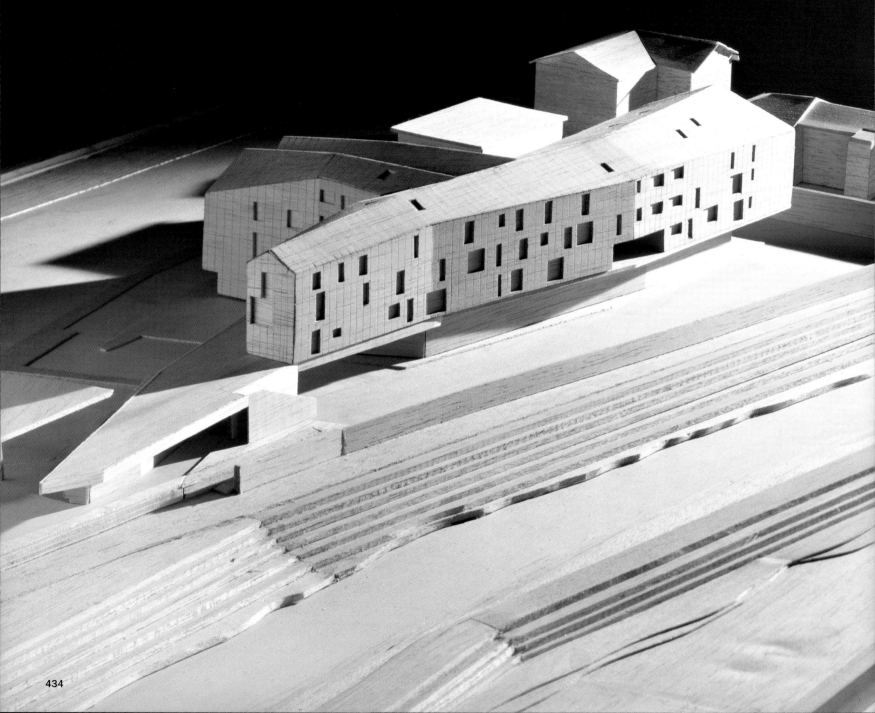

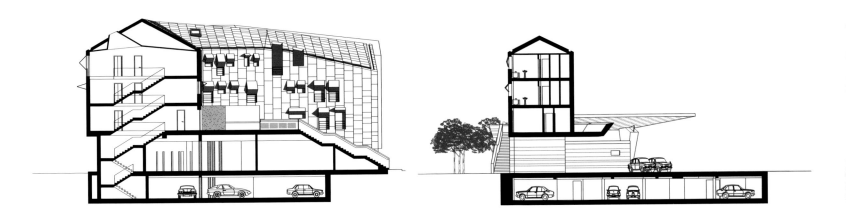

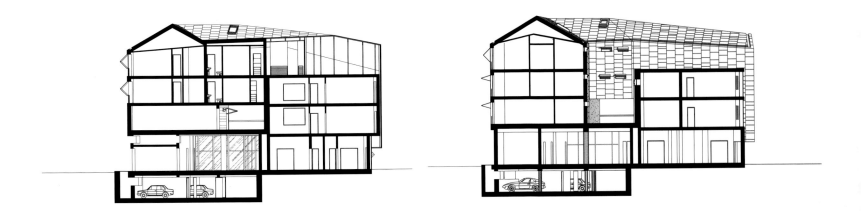

cross sections
横截面

0   5 m

439

MERATE TOWN HALL

Merate, Lecco, Italy, 2001 - under construction

地点	Merate，莱科，意大利
项目	市政厅和礼堂
客户	Merate市政府
结构	Studio Myallonnier
系统	Studio Armondi
规划	2001年-2005年
建设	在建
成本	3,615,200欧元
建筑面积	4,725平方米
体量	21,700立方米
承包商	Giacomo Zenga Impianti Tecnologici Sbrescia s.n.c.

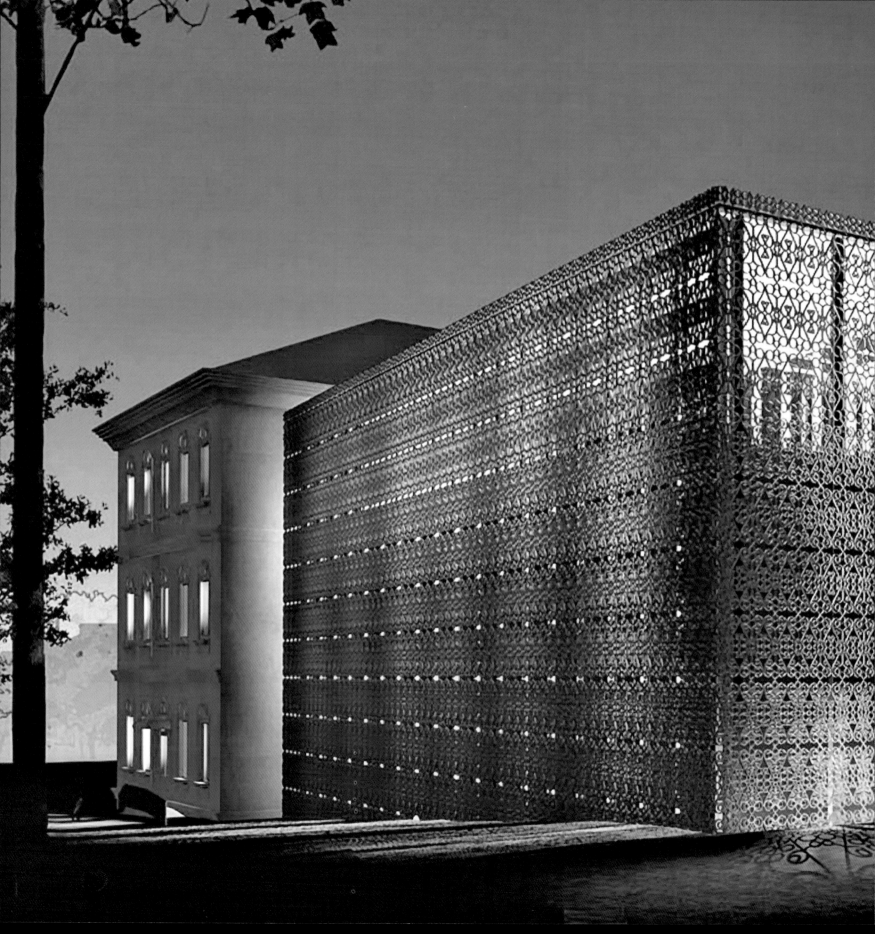

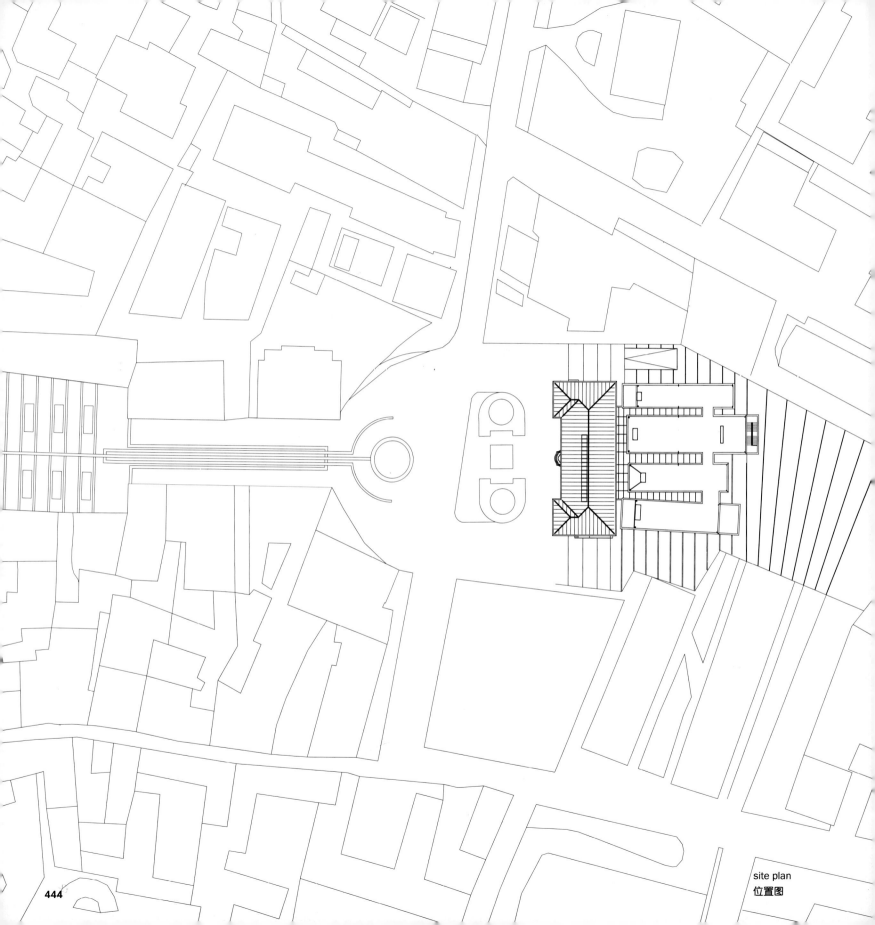

site plan
位置图

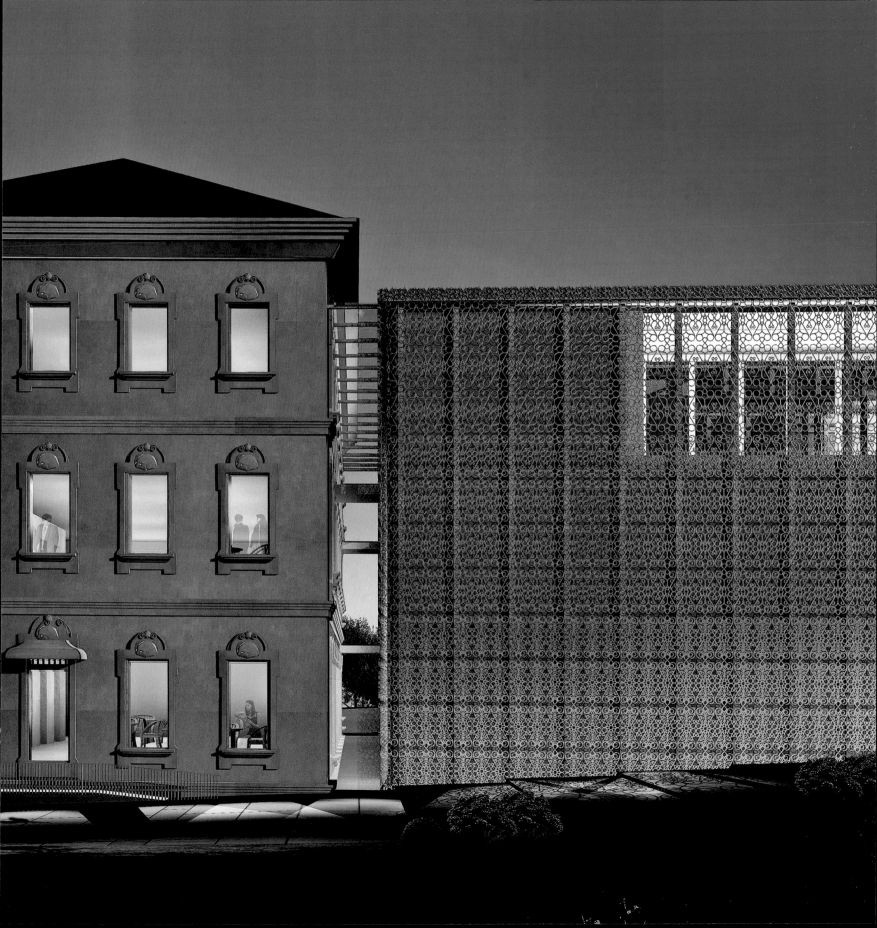

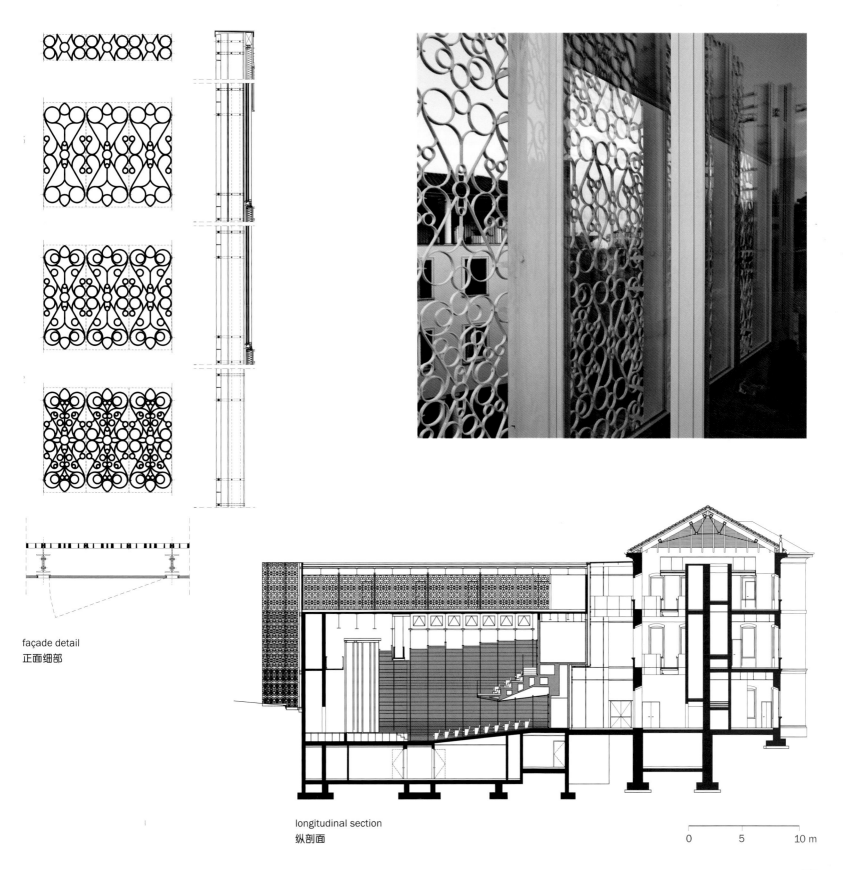

façade detail
正面细部

longitudinal section
纵剖面

0   5   10 m

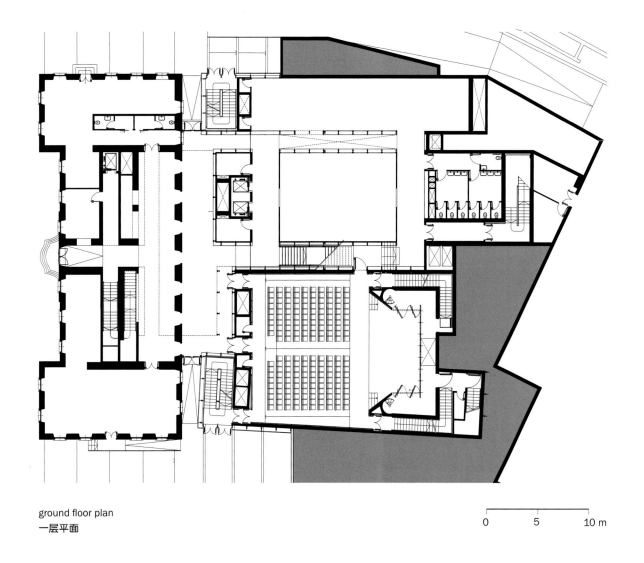

ground floor plan
一层平面

0   5   10 m

PORTA SUSA STATION

Turin, Italy, 2001

地点	都灵，意大利
项目	基础设施
客户	国家铁路局，运输和运行服务公司，基础设施分部
结构	Studio Gambogi
规划	2001年设计竞赛

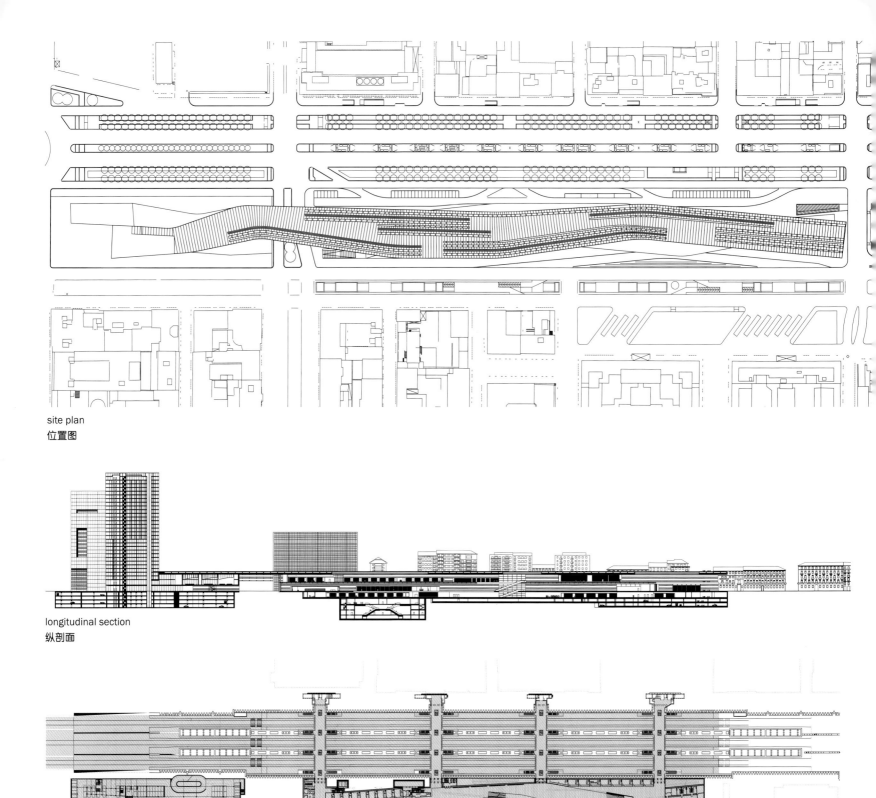

site plan
位置图

longitudinal section
纵剖面

plan level -3.20
高度-3.20平面

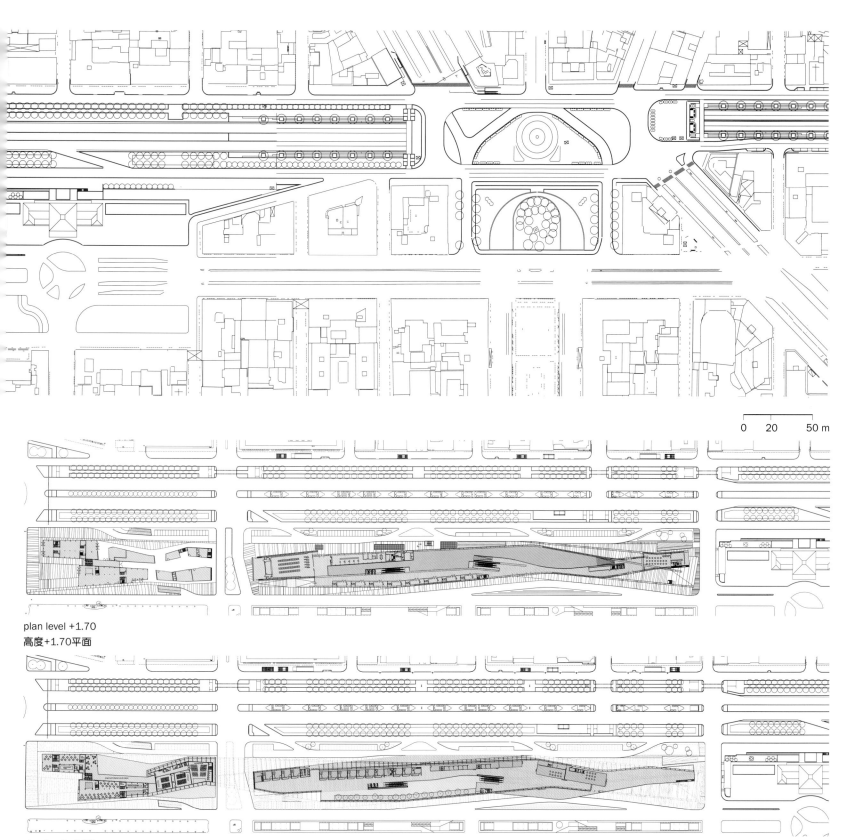

0 20 50 m

plan level +1.70
高度+1.70平面

plan level +9.20 m
高度+9.20米平面

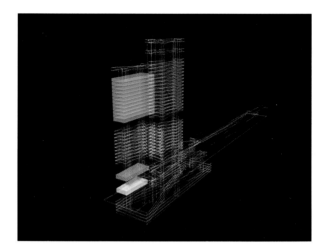

funtional scheme
功能设计

⬡ entrance hall / 入口大厅
⬤ restaurant / 餐厅
⬤ three stars hotel / 三星酒店

POOL IN GAZZANIGA

Gazzaniga, Bergamo, Italy, 2001-2002

地点 Gazzaniga，贝加莫，意大利
项目 住宅
客户 私人客户
规划 2001年
建设 2002年
成本 800,000欧元

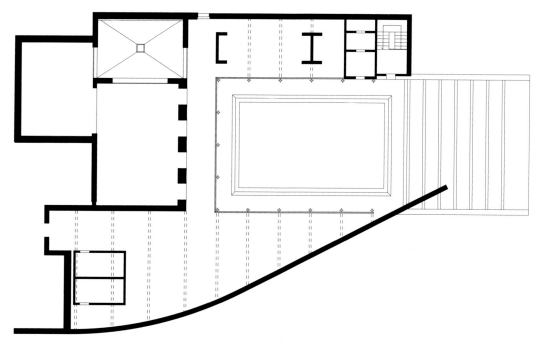

ground floor plan
一层平面

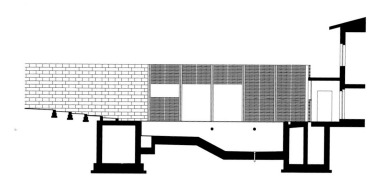

cross section
横剖面

0 1   5 m

EX AREA FIAT
RESIDENTIAL COMPLEX

Florence, Italy, 2001 - in progress

地点　佛罗伦萨，意大利
项目　住宅建筑群
客户　Immobiliare Novoli S.p.A.
规划　2001年
建设　在建
成本　6,000,000欧元
建筑面积　4,500平方米
体量　14,000立方米

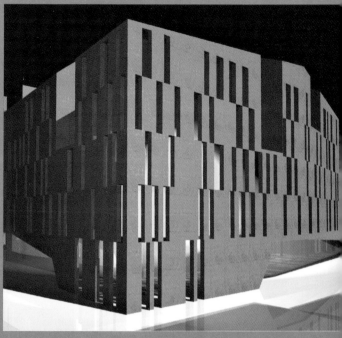
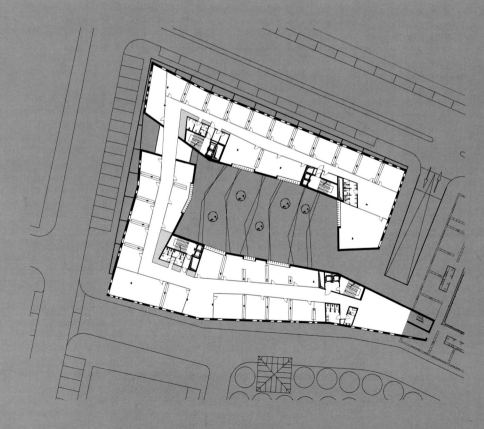

first floor plan
二层平面

0   5   10 m

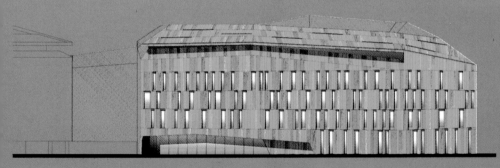

north elevation
北面正立图

east elevation
东面正立图

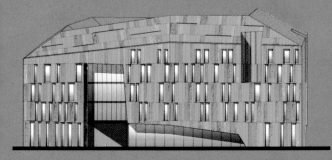

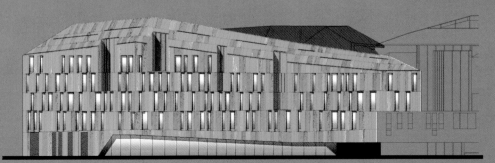

west elevation
西面正立图

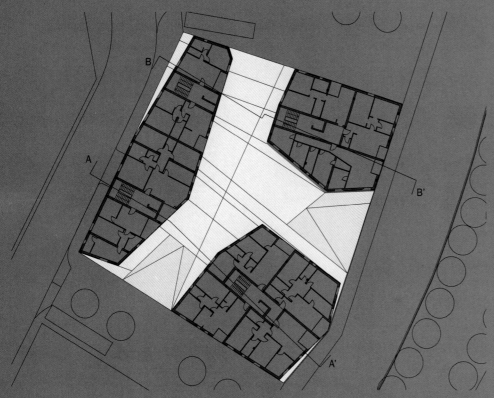

type plan
类型规划

0   5 m

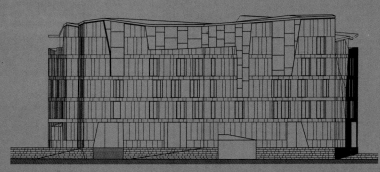

west elevation
西立面

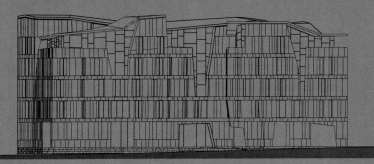

east elevation
东立面

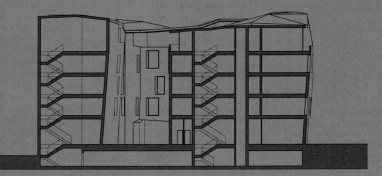

section A-A'
A-A'剖面

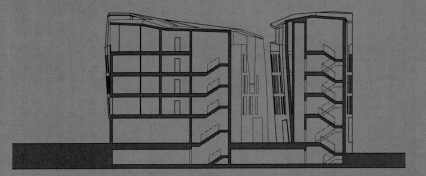

section B-B'
B-B'剖面

0 5 m

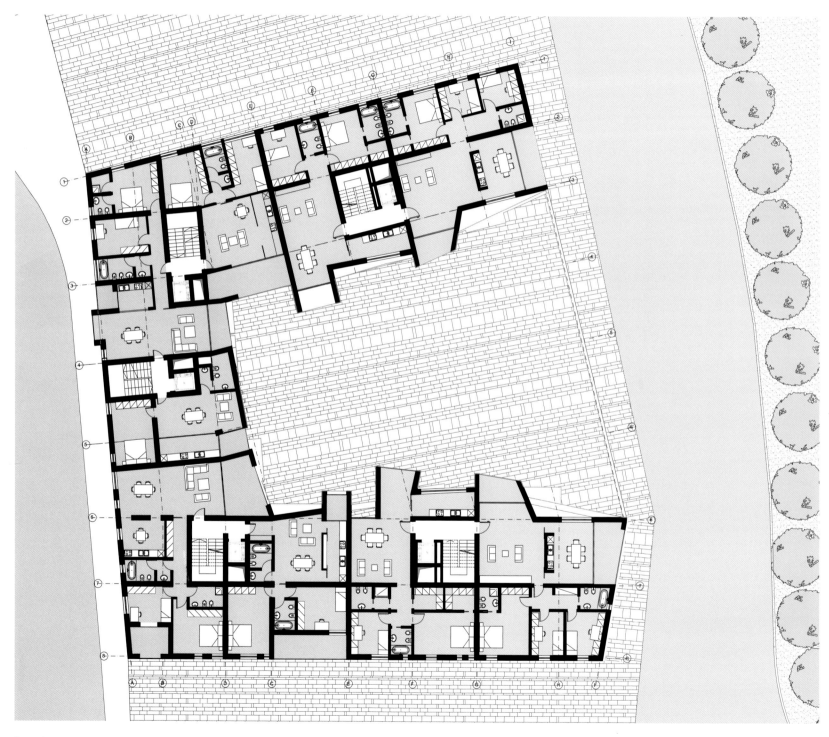

type plan
平面类型

0 1   5 m

elevation
正立面

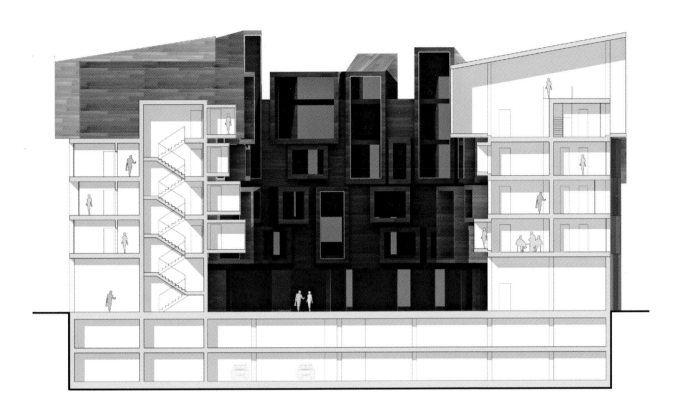

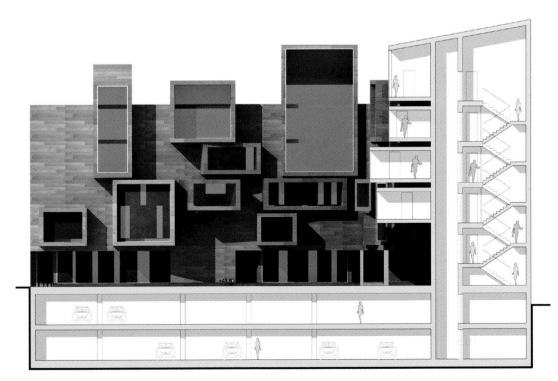

sections
剖面图　　0　　　5 m

NEMBRO LIBRARY

Nembro, Bergamo, Italy, 2002-2007

地点　　Nembro，贝加莫，意大利
项目　　图书馆和文化中心
客户　　Nembro市政府
结构　　Favero&Milan Ingegneria
温度控制系统公司
事务所　Tecnico Zambonin
电气系统　Eros Grava
规划　　2002年
施工　　2005年-2007年
成本　　1,888,250欧元
建筑面积　1,875平方米
体量　　11,200立方米
承包商　Zeral S.r.l. Costruzioni edili

site plan
位置图

first floor plan
二层平面

0   5   10 m

494

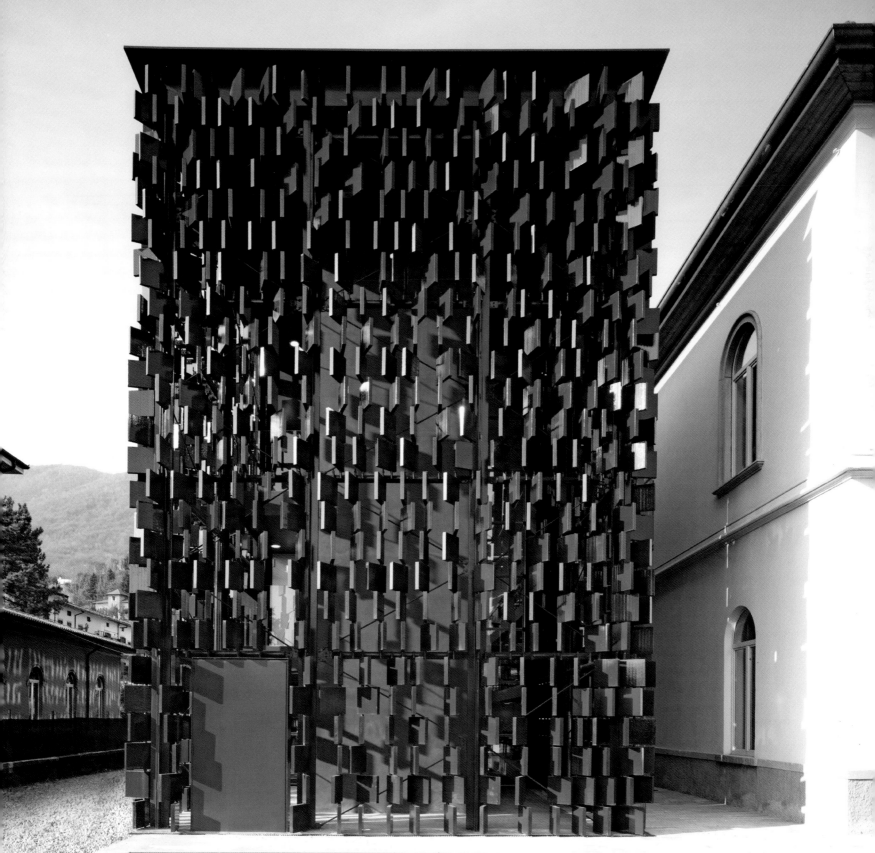

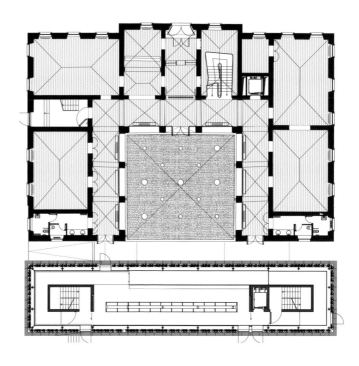

ground floor plan
一层平面

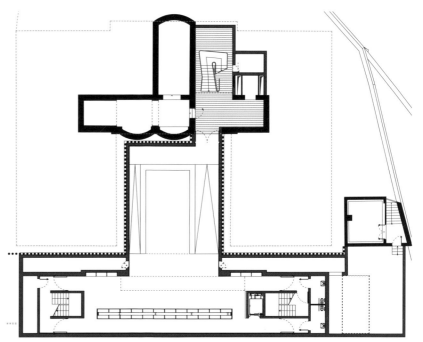

plan level -3.30
水平平面图-3.30

0   5   10 m

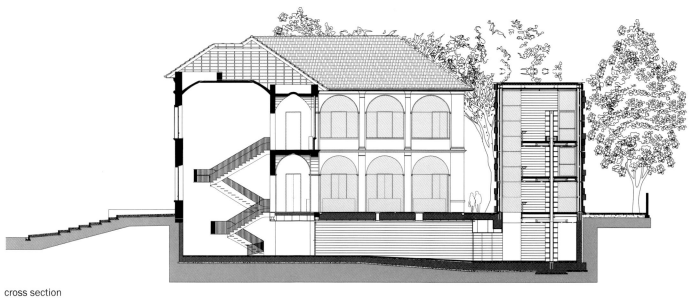

cross section
横剖面

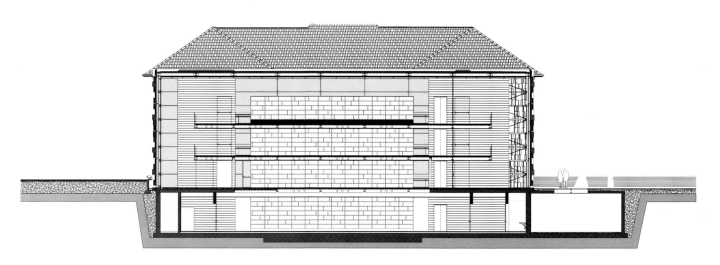

longitudinal section
纵剖面

0   5   10 m

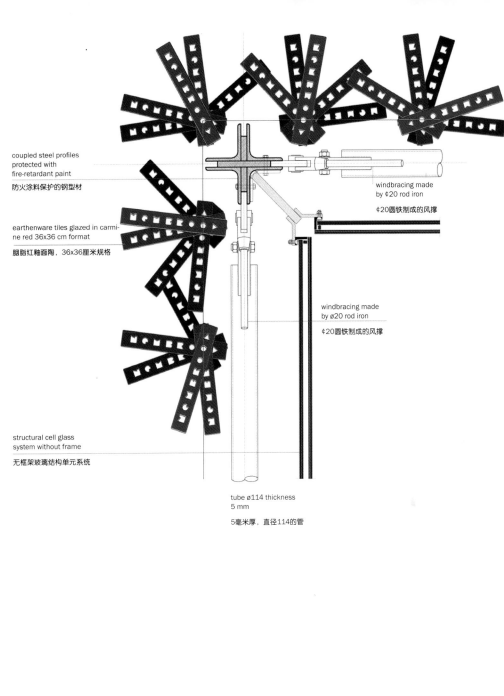

coupled steel profiles
protected with
fire-retardant paint

防火涂料保护的钢型材

earthenware tiles glazed in carmine red 36x36 cm format

胭脂红釉面陶，36x36厘米规格

windbracing made
by ⊄20 rod iron

⊄20圆铁制成的风撑

windbracing made
by ø20 rod iron

⊄20圆铁制成的风撑

structural cell glass
system without frame

无框架玻璃结构单元系统

tube ø114 thickness
5 mm

5毫米厚，直径114的管

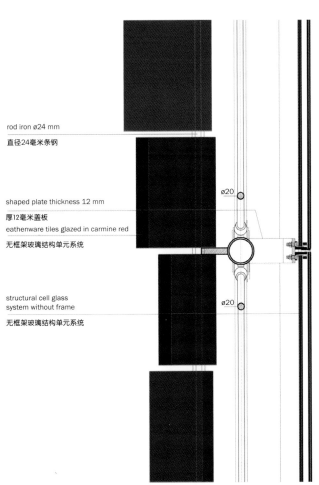

rod iron ø24 mm

直径24毫米条钢

shaped plate thickness 12 mm

厚12毫米盖板

eathenware tiles glazed in carmine red

无框架玻璃结构单元系统

structural cell glass
system without frame

无框架玻璃结构单元系统

ø20

ø20

corner detail

转角细部

0   20 cm

CDD-CENTER FOR DISABILITY

Seregno, Milan, Italy
2003 - under construction

地点　Seregno，米兰，意大利
项目　学校
客户　Seregno市政府
结构　Matteo Fiori and Luca Varesi
系统　Studio Armondi - StudioTi
规划　2003年
建设　在建
成本　2,700,000欧元
占地面积　10,000平方米
建筑面积　1,300平方米
体量　4,940立方米
承包商　T.i.e.c.i. S.r.l.

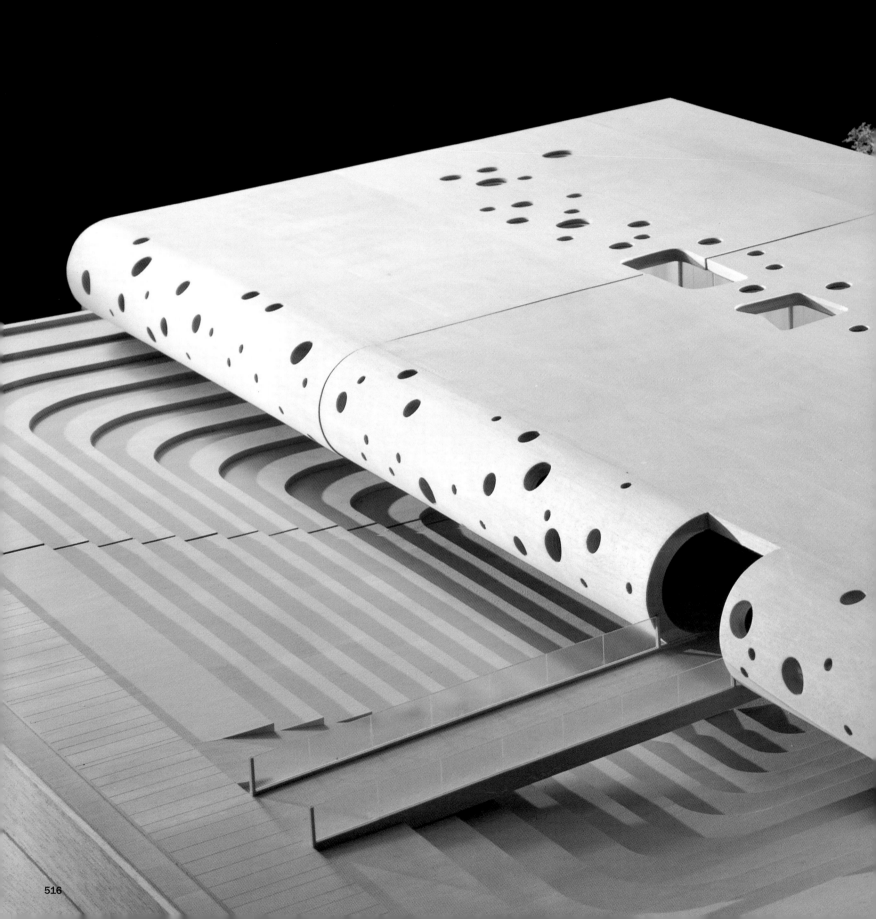

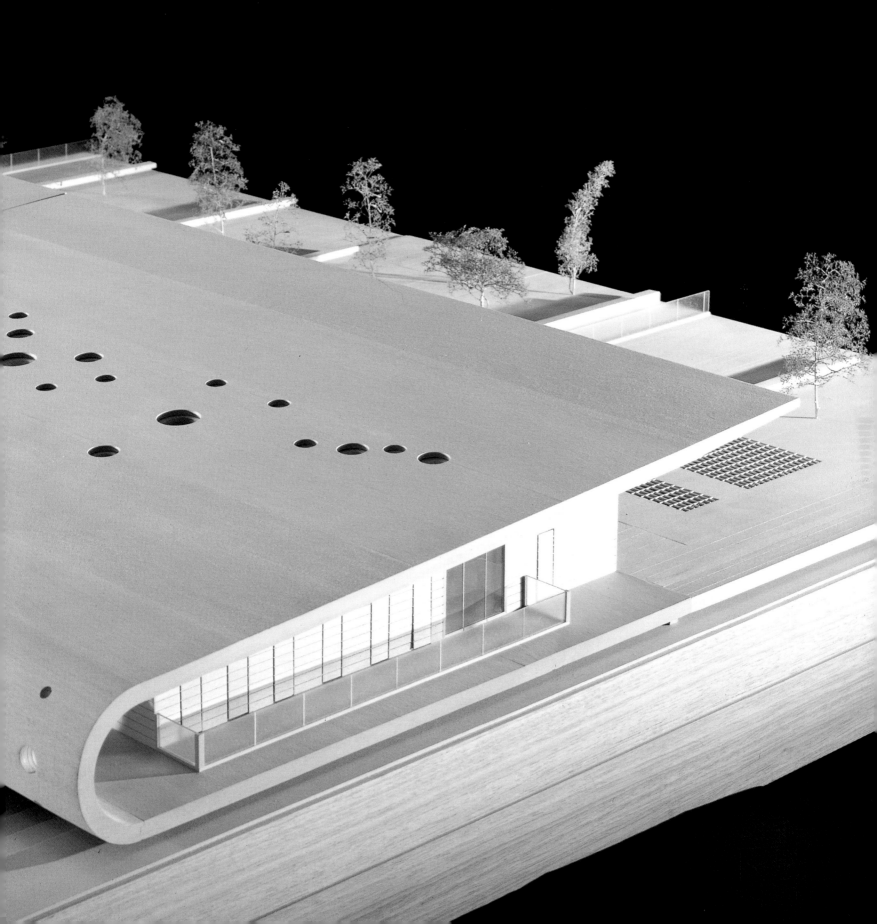

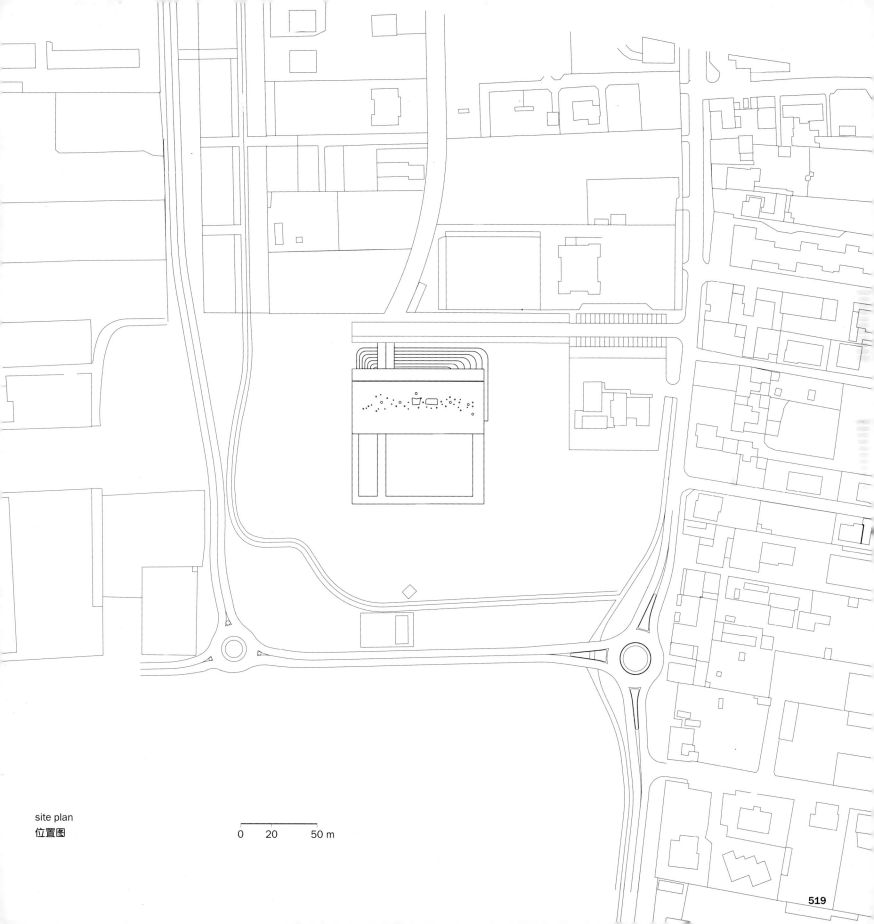

site plan
位置图

0   20   50 m

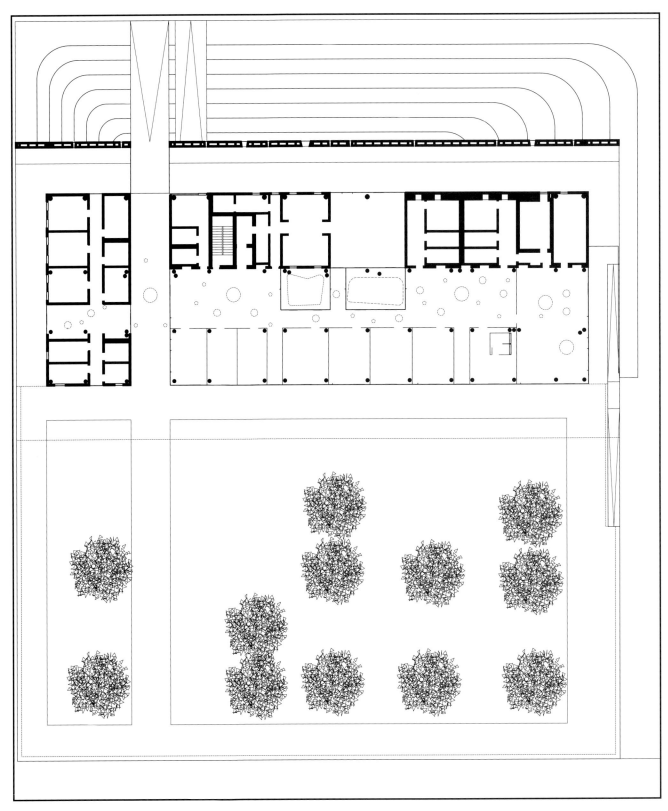

ground floor plan
一层规划

0　　　5　　　10 m

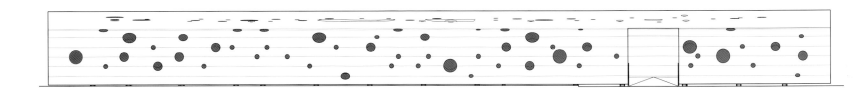

longitudinal elevation

垂直正视图

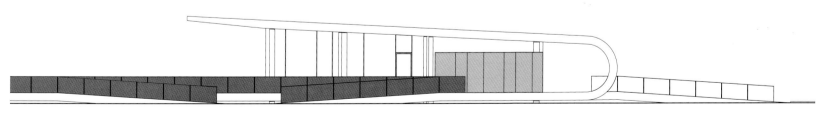

cross elevation
横立面

0   5 m

THE CORD

entrance to the 50th Venice Biennial of Visual Art
Giardini di Castello, Venice, Italy, 2003

地点　Giardini di Castello，威尼斯，意大利
项目　第50届威尼斯双年展视觉艺术展入口
客户　Biennale di Venezia
结构　Favero&Milan Ingegneria
照明设计　Jan Van Lierde
规划　2003年
建设　2003年
成本　600,000欧元
占地面积　250平方米
体量　1,200立方米
承包商　Fima Cosma Silos S.r.l.

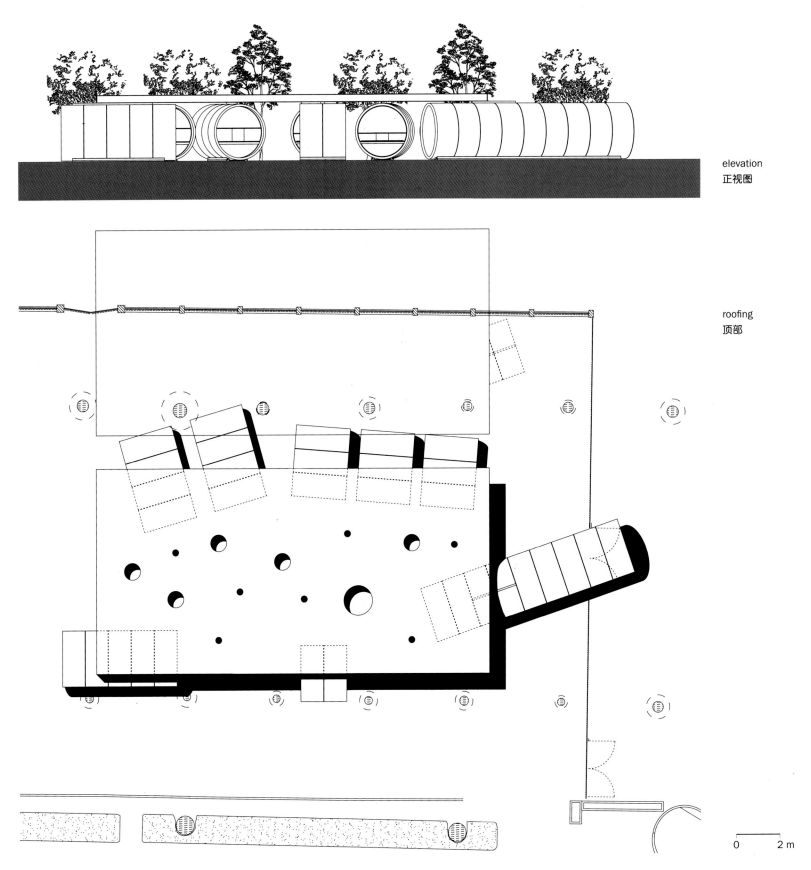

elevation
正视图

roofing
顶部

0 2 m

Yan Lei and Fu J...
Jean-Paul Jungmann and Tamas...
Sora Kim and Gi...
Rem Koolhaas   Gabriel Kur...

Din Q. Le   ...
Kami...
Sarah Lucas   M/M   A...

Cildo Meireles
Helen Mirra   Andrei Monastirsky
Rabih Mroué   Sabah Naim   Delman...
Jun Nguyen-Hatsushiba
Roman Ondák   Yoko Ono
Olumuyiwa Olamide Osifuye
Jennifer Pastor   Yan Pei-M...
Manfred Pernice   Diego Perrone
Magnus von Plessen
Jorge Queiroz   Wal...
Charles R...
Pia Roenike   Fernando
Markus Schinwal...
Wael Shawky
Andreas Slominski   P...
Monika Sonowa...
Superflex

Tatian Trouvé
Jaya...
Lew...
Amelie...

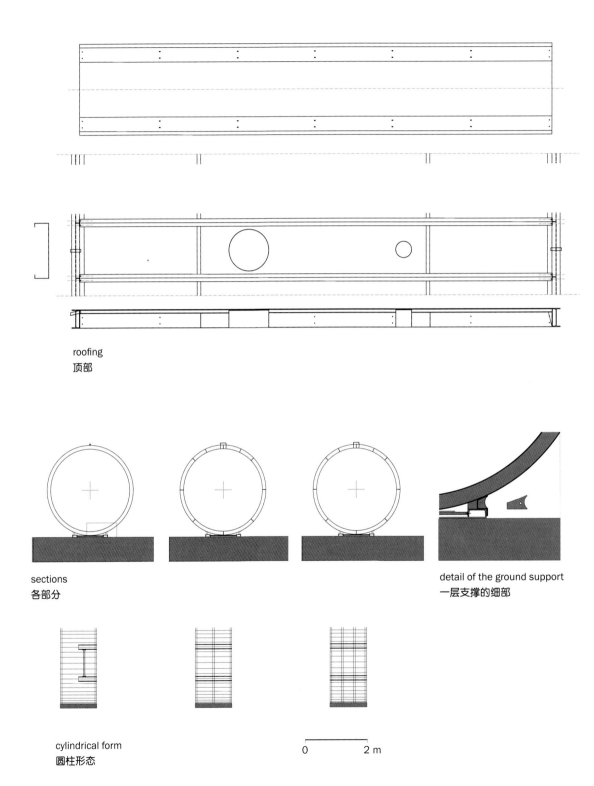

roofing
顶部

sections
各部分

detail of the ground support
一层支撑的细部

cylindrical form
圆柱形态

0   2 m

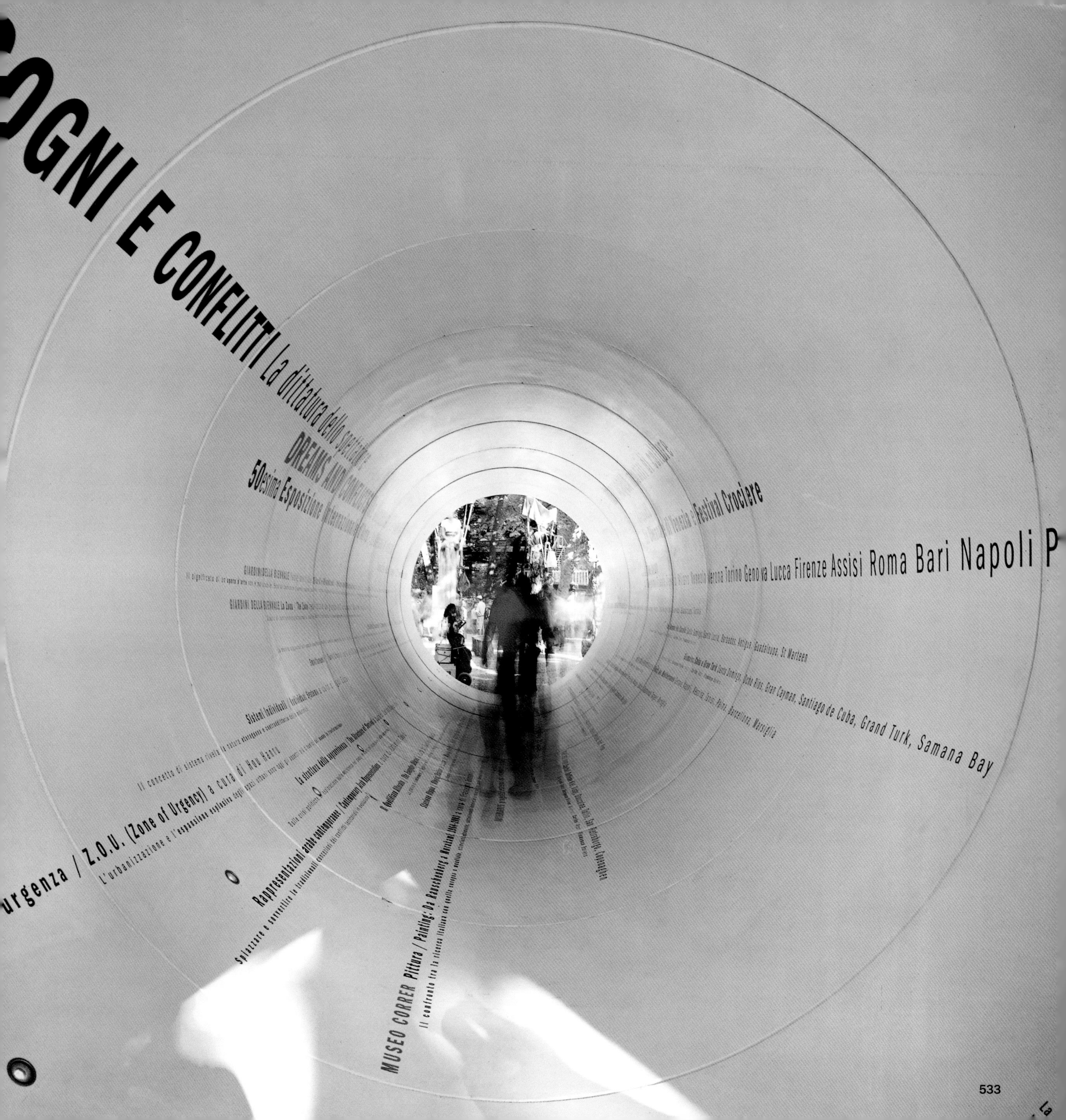
OGNI E CONFLITTI La dittatura dello
DREAMS AND
50esima Esposizione
Festival Crociere
Verona Torino Geno va Lucca Firenze Assisi Roma Bari Napoli P
Barbados, Antigua, Guadeloupe, St Marteen
Santo Domingo, Ocho Rios, Gran Cayman, Santiago de Cuba, Grand Turk, Samana Bay
GIARDINI DELLA BIENNALE La Zona / The Zone
Sistemi individuali / Individual Systems
Il concetto di sistema rivela la natura
urgenza / Z.O.U. (Zone of Urgency) a cura di Hou Hanru
L'urbanizzazione e l'espansione esplosiva
La struttura della sopravvivenza / The Structure of Survival
Rappresentazioni arabe contemporanee / Contemporary Arab Representation
MUSEO CORRER Pittura / Painting: Da Rauschenberg a Murakami 1964-2003
Il confronto tra la ricerca italiana
La

SOGNI E CONFLIT
DREAMS AND
50esima Esposizione I

GIARDINI DELLA BIENNALE Padiglione Italia Rita

Il significato di un'**opera d'arte** non è mai stab

GIARDINI DELLA BI

Smottamenti

Arte africana contemporanea e paesaggi i

el Systems a cura di

TI La dittatura

CONFLICTS Th

ernazionale d'Art

Ve

e Rivoluzioni / **Delays and Revolutions** a

sso al contrario dipende spesso da nuove letture e traduzi

ALE **La Zona / The Zone** realizzato da

di confronto e piattaforma di dialogo dove si sperimenta

ARSENALE Clandestini

a mostra suggerisce uno **spazio condiviso** privo di confin

ault Lines a cura di Gilane Tawad

biamento gli smottamenti segnano rigetti significativ

or Zabel

modernità

ou Hanru

i aspetti più dinamici

535

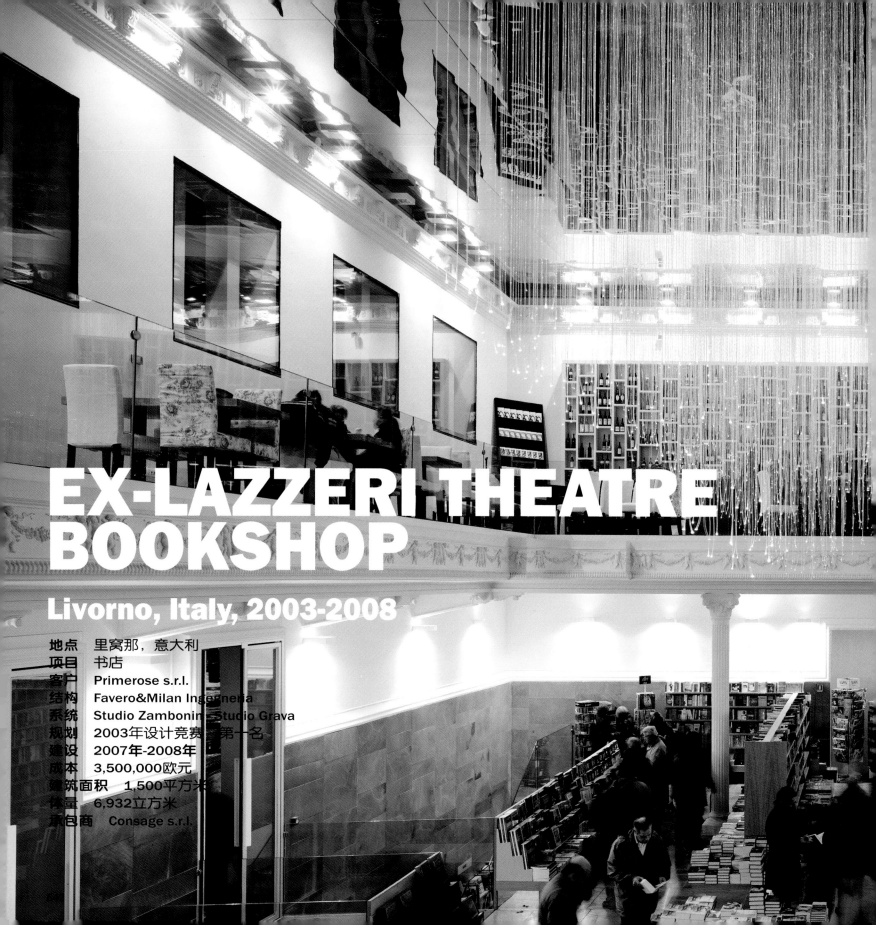

EX-LAZZERI THEATRE
BOOKSHOP

Livorno, Italy, 2003-2008

地点　里窝那，意大利
项目　书店
客户　Primerose s.r.l.
结构　Favero&Milan Ingegneria
系统　Studio Zambonin, Studio Grava
规划　2003年设计竞赛　第一名
建设　2007年-2008年
成本　3,500,000欧元
建筑面积　1,500平方米
体量　6,932立方米
承包商　Consage s.r.l.

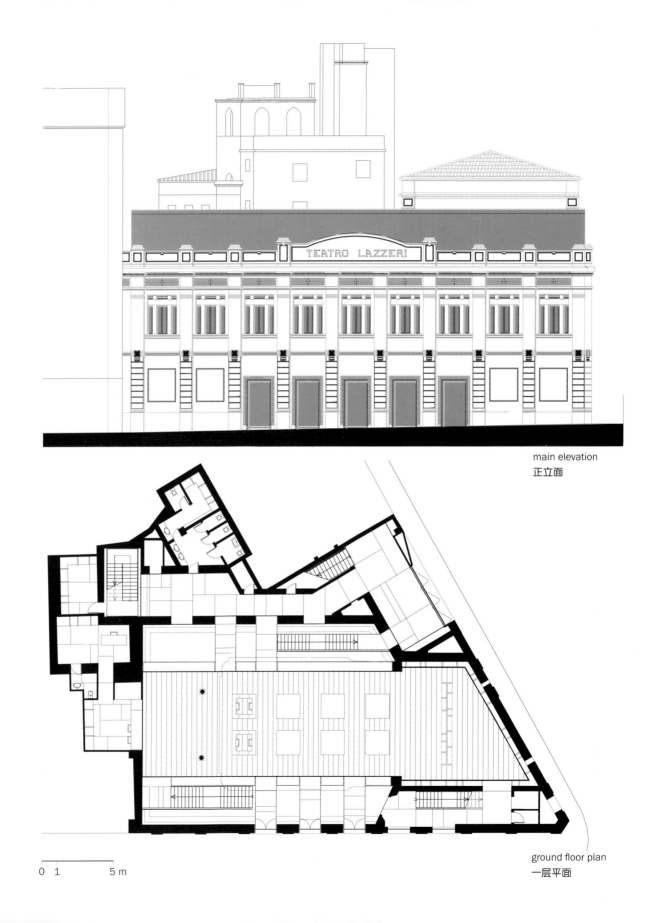

main elevation
正立面

ground floor plan
一层平面

0 1   5 m

interior of Lazzeri cinema-theatre before the renovation project
改建前的拉扎列影剧院的内部状况

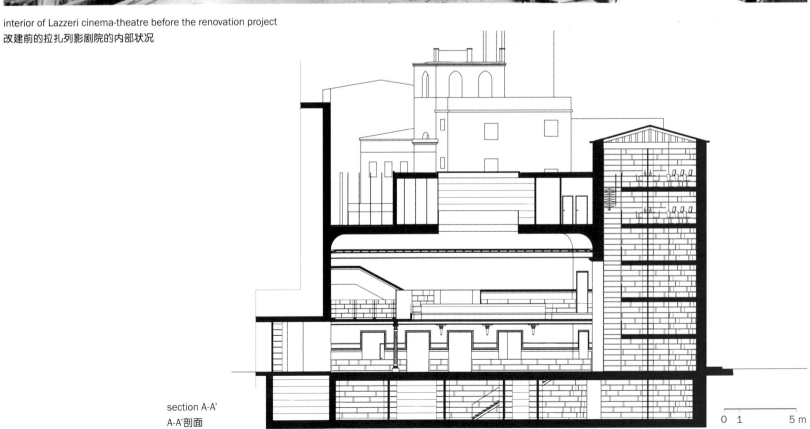

section A-A'
A-A'剖面

0 1 5 m

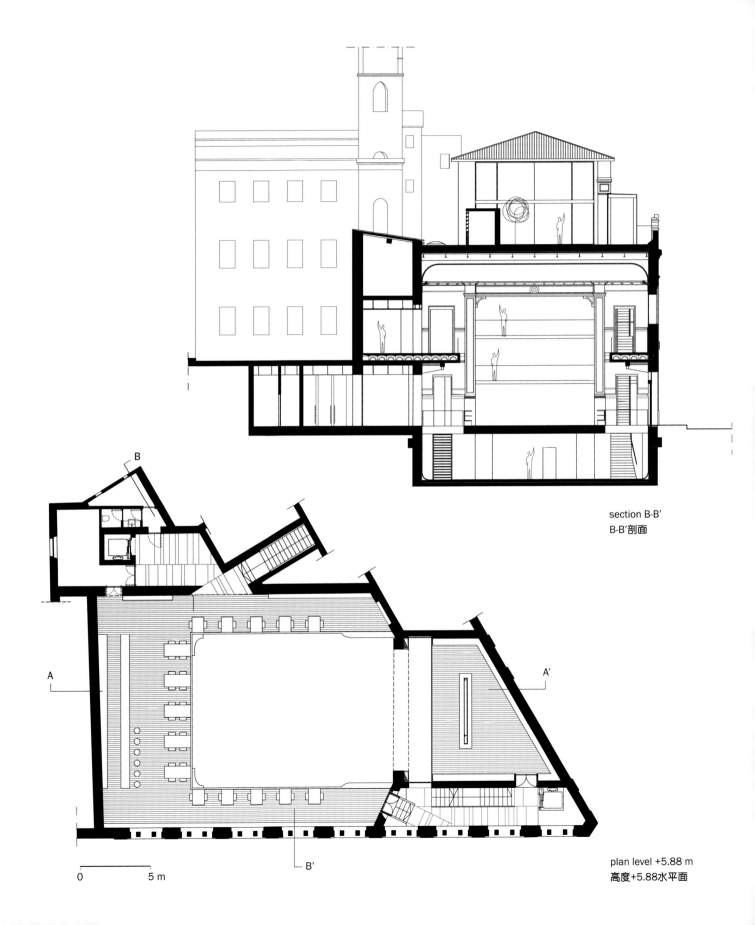

section B-B'
B-B'剖面

plan level +5.88 m
高度+5.88水平面

0   5 m

NEW MULTIPURPOSE CENTRE

Trieste, Italy, 2004 - under construction

地点　的里雅斯特，意大利
项目　多功能中心
客户　Fondazione CRTrieste
结构　Favero&Milan Ingegneria
系统　StudioTi
规划　2004年
建设　在建
成本　15,000,000欧元
建筑面积　3,600平方米
体量　11,000立方米

axonometric view
轴测视图

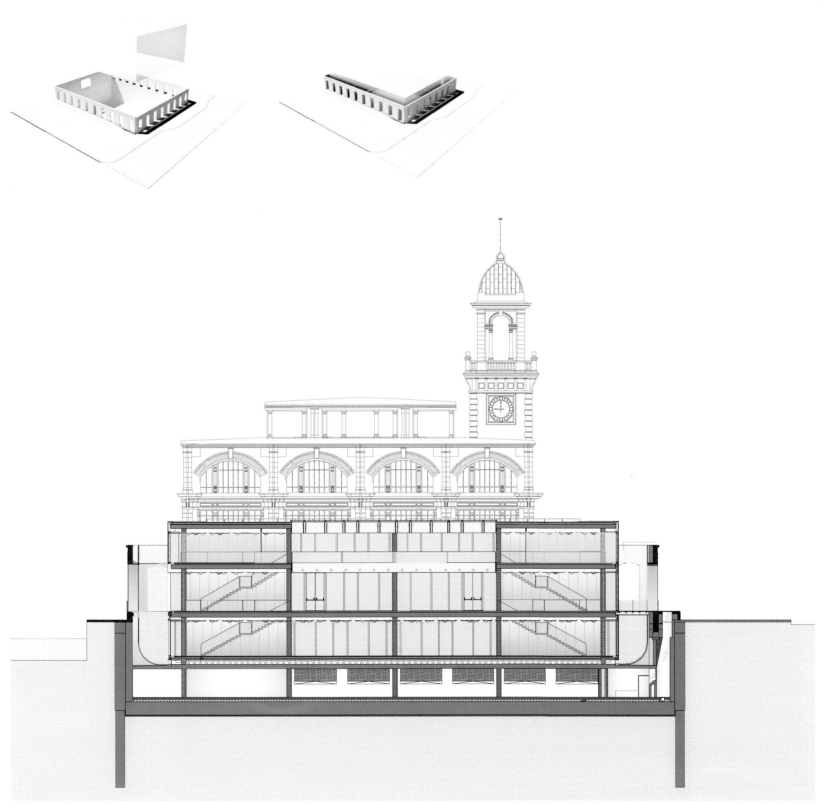

longitudinal section
纵剖面

0   5   10 m

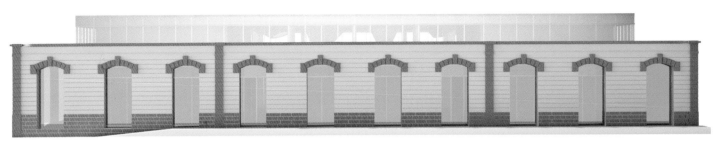

east elevation
东面正视图

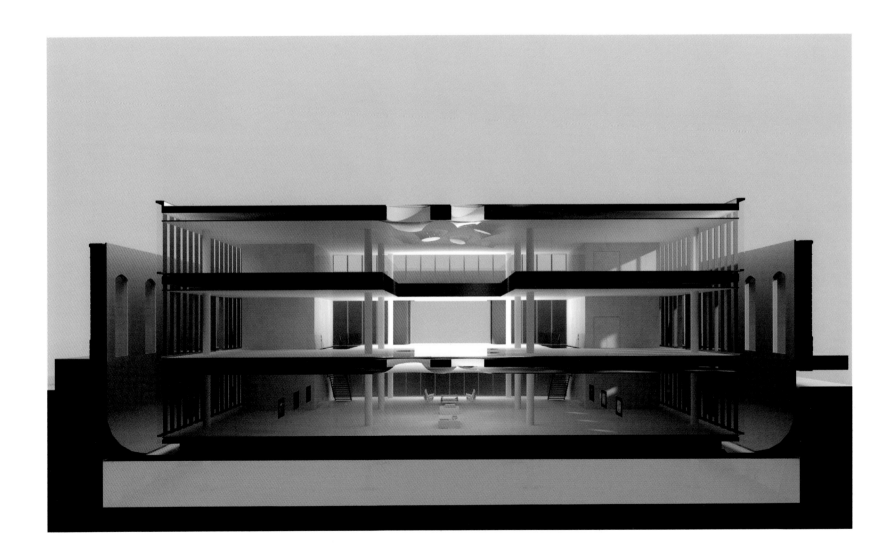

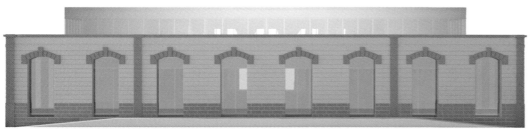

south elevation
南面正视图

0   1   5 m

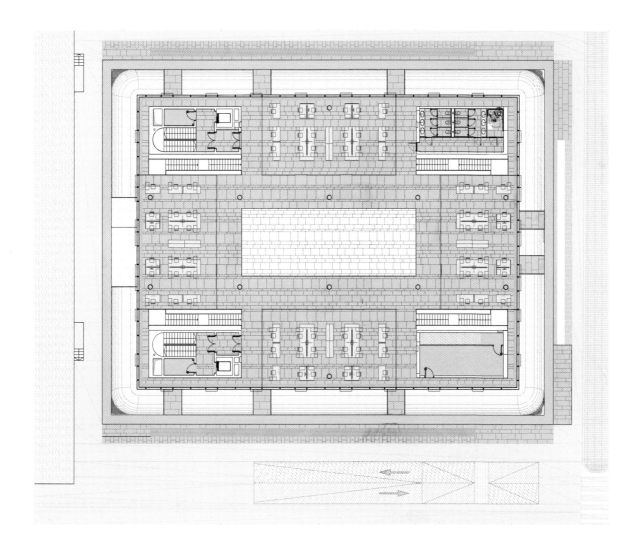

plan level +5.20 m
高度+5.20水平米

0   5   10 m

PERFETTI VAN MELLE FACTORY RENOVATION

Lainate, Milan, Italy, 2005-2011

地点　Lainate，米兰，意大利
项目　办公楼和仓库
客户　Perfetti Van Melle S.p.A.
结构　Favero&Milan Ingegneria
系统　StudioTi
规划　2005年
建设　2007年-2011年
成本　15,000,000欧元
建筑面积　12,985平方米
体量　94,571立方米
承包商　Rizzani De Eccher

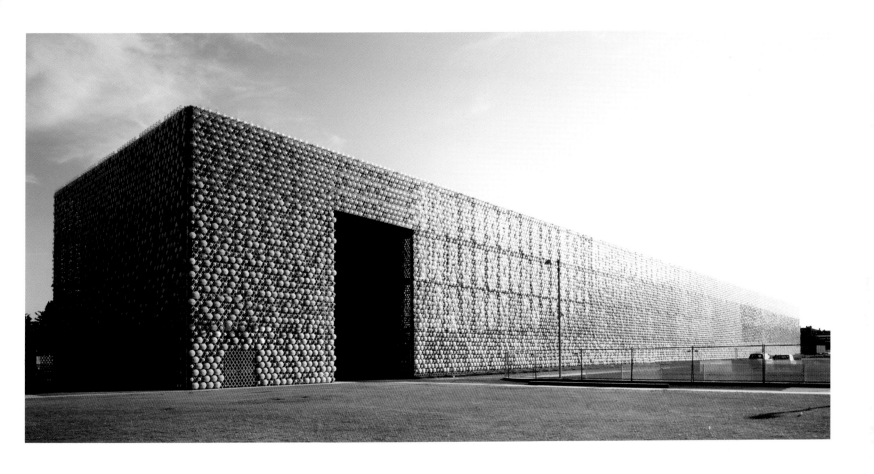

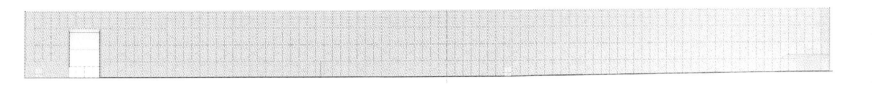

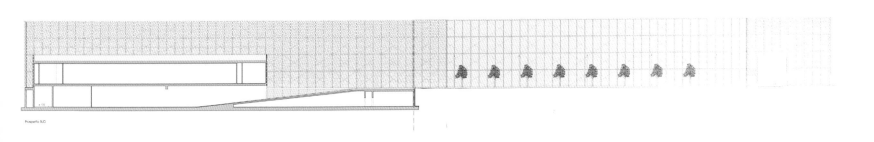

longitudinal elevation

垂直正视图

0   5   20 m

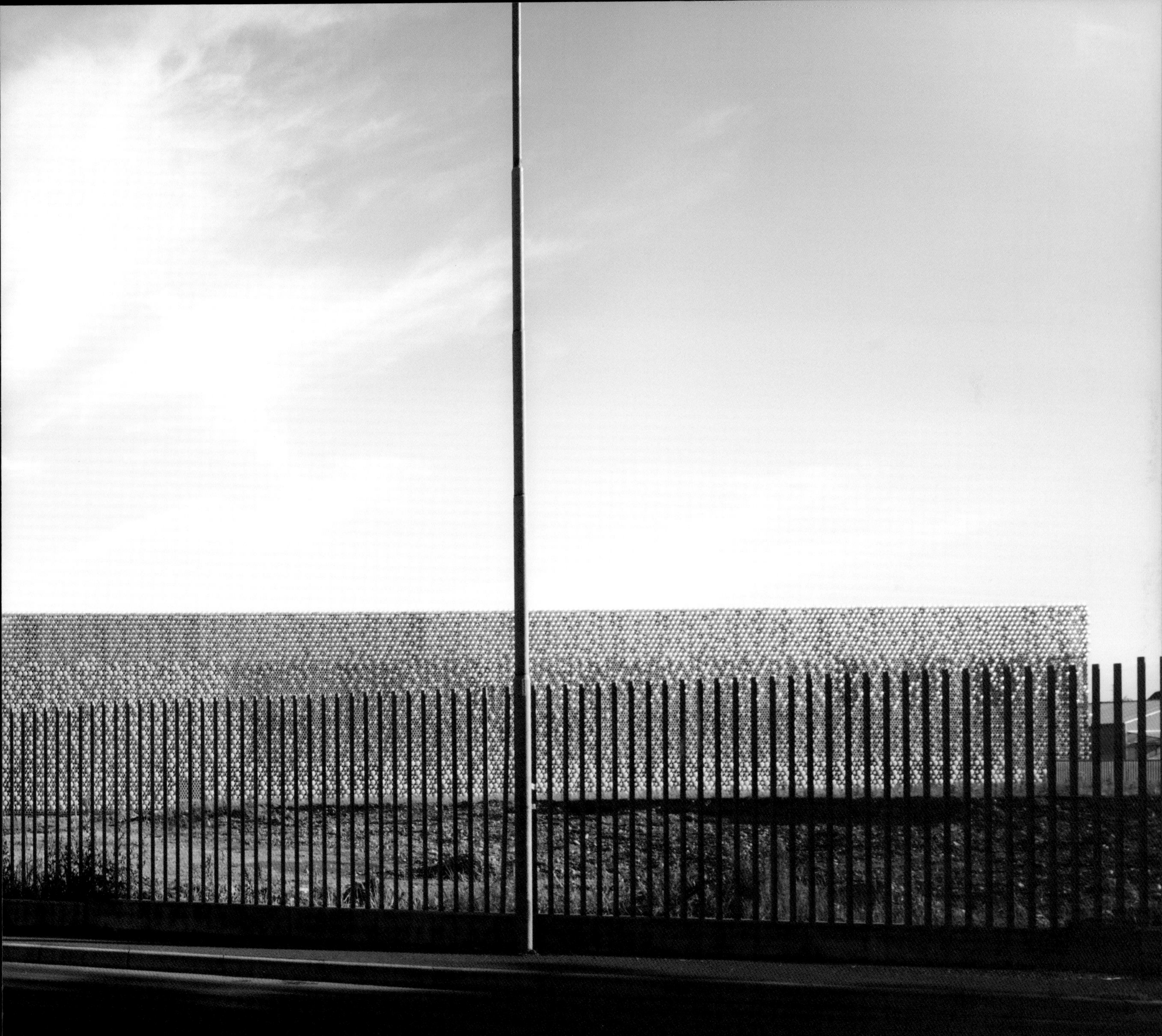

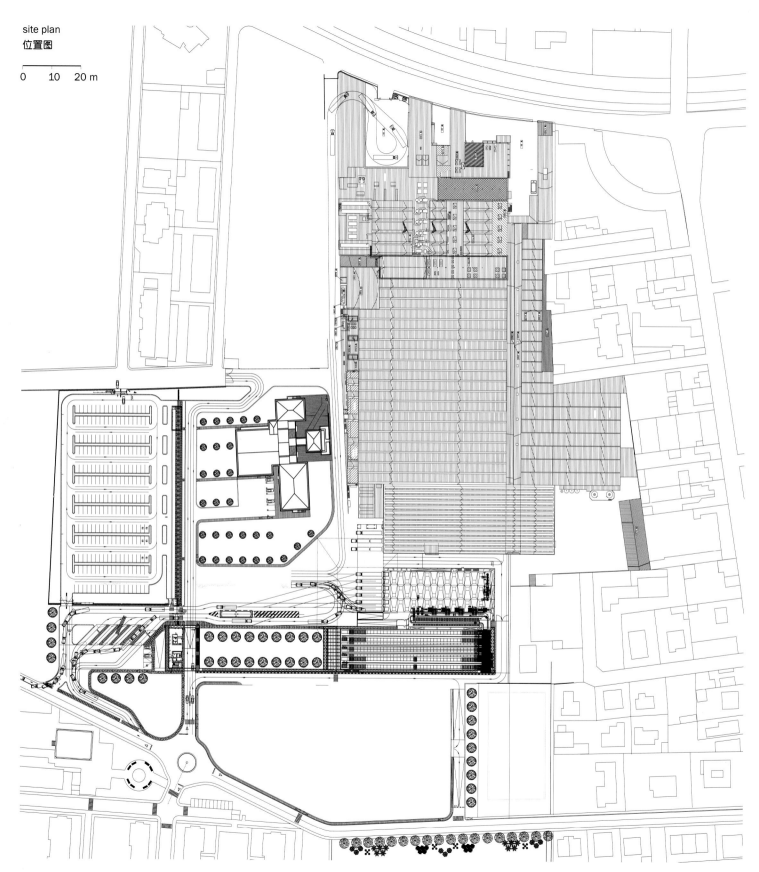

site plan
位置图

0   10   20 m

texture of the façade
正面纹理

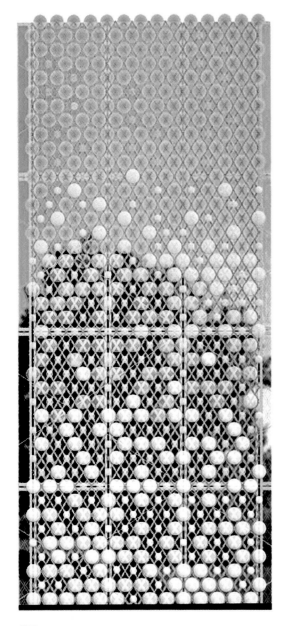
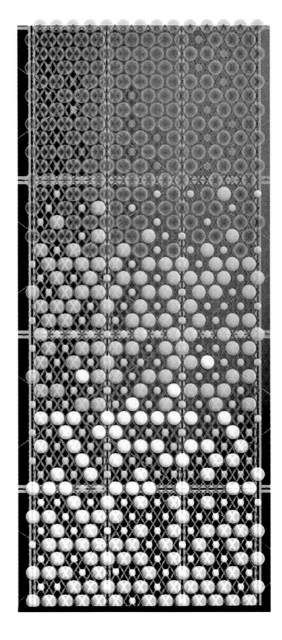
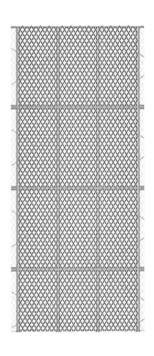

4 EVER GREEN

multipurpose tower
Tirana, Albania, 2005 - under construction

地点	地拉那，阿尔巴尼亚
项目	多功能塔楼
客户	Al&Gi Shipk RR - Dervish Hima
结构	aei progetti - Niccolò De Robertis
系统	StudioTi
规划	2005年设计竞赛，第一名
建设	在建
成本	25,000,000欧元
建筑面积	12,400平方米
体量	46,750立方米

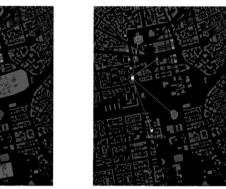

preliminary urban studies
城区初步研究

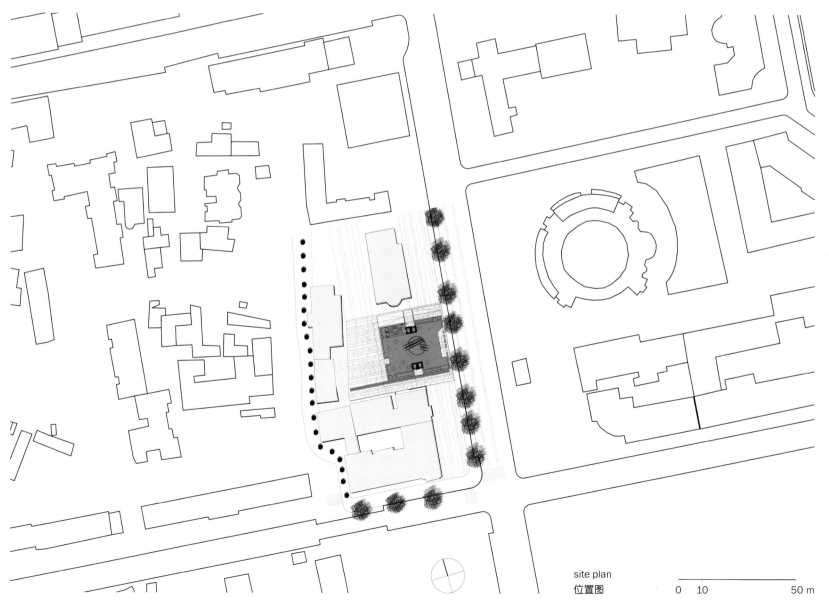

site plan
位置图

0   10   50 m

585

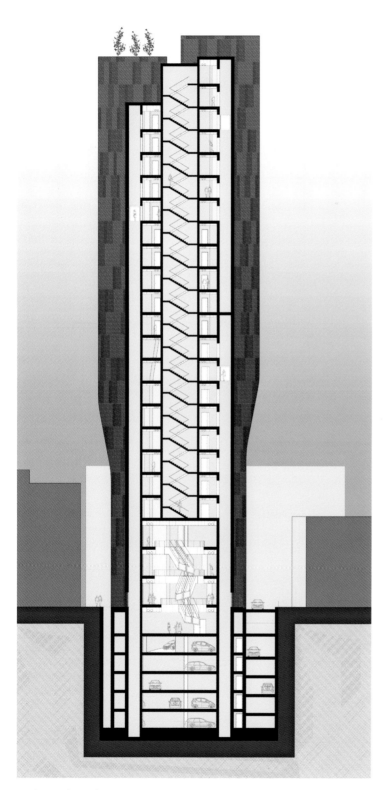

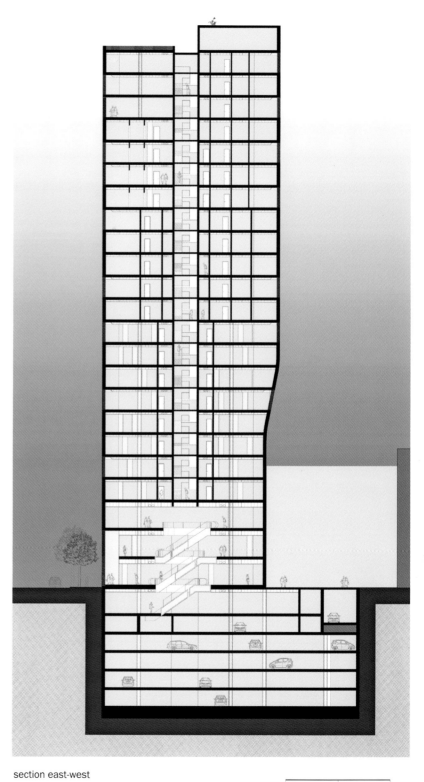

section north-south
南北剖面

section east-west
东西剖面

0   5   10 m

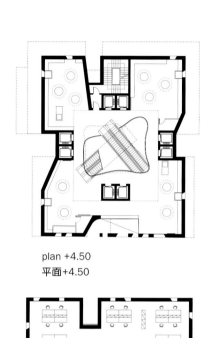

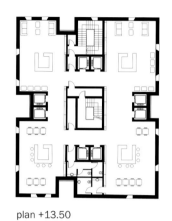

plan +4.50
平面+4.50

plan +9.00
平面+9.00

plan +13.50
平面+13.50

plan +31.80
平面+31.80

plan +42.15
平面+42.15

plan +42.15
平面+42.15

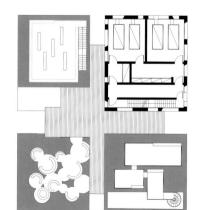

plan +73.20
平面+73.20

plan +76.55
平面+76.55

plan +80.10
平面+80.10

0   5   10 m

east elevation
东面正视图

south elevation
南面正视图

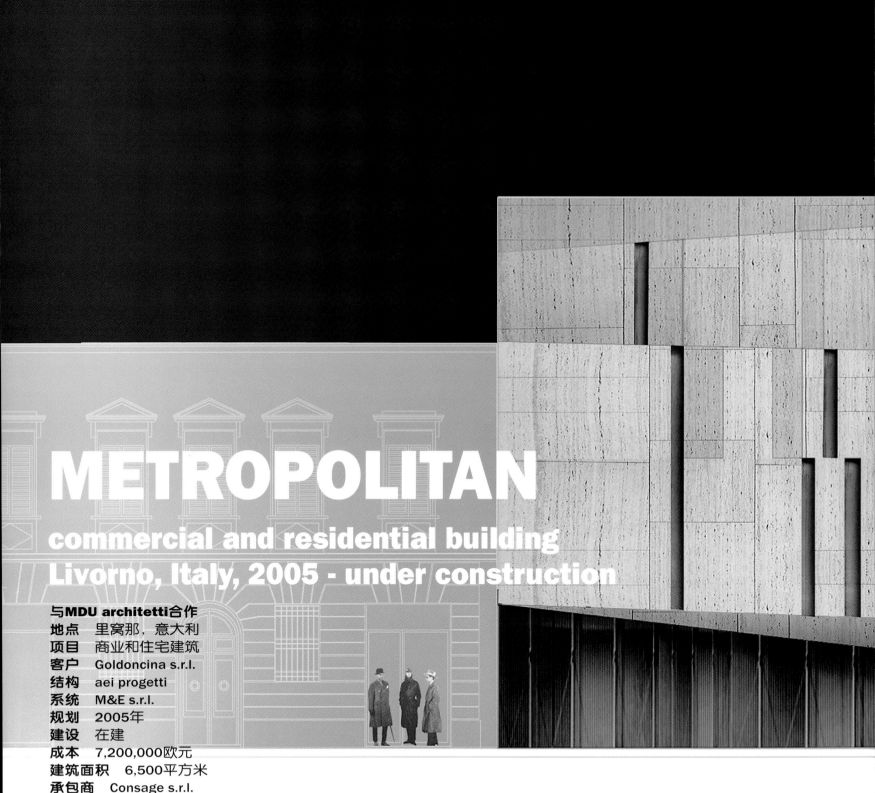

METROPOLITAN
commercial and residential building
Livorno, Italy, 2005 - under construction

与MDU architetti合作
地点 里窝那，意大利
项目 商业和住宅建筑
客户 Goldoncina s.r.l.
结构 aei progetti
系统 M&E s.r.l.
规划 2005年
建设 在建
成本 7,200,000欧元
建筑面积 6,500平方米
承包商 Consage s.r.l.

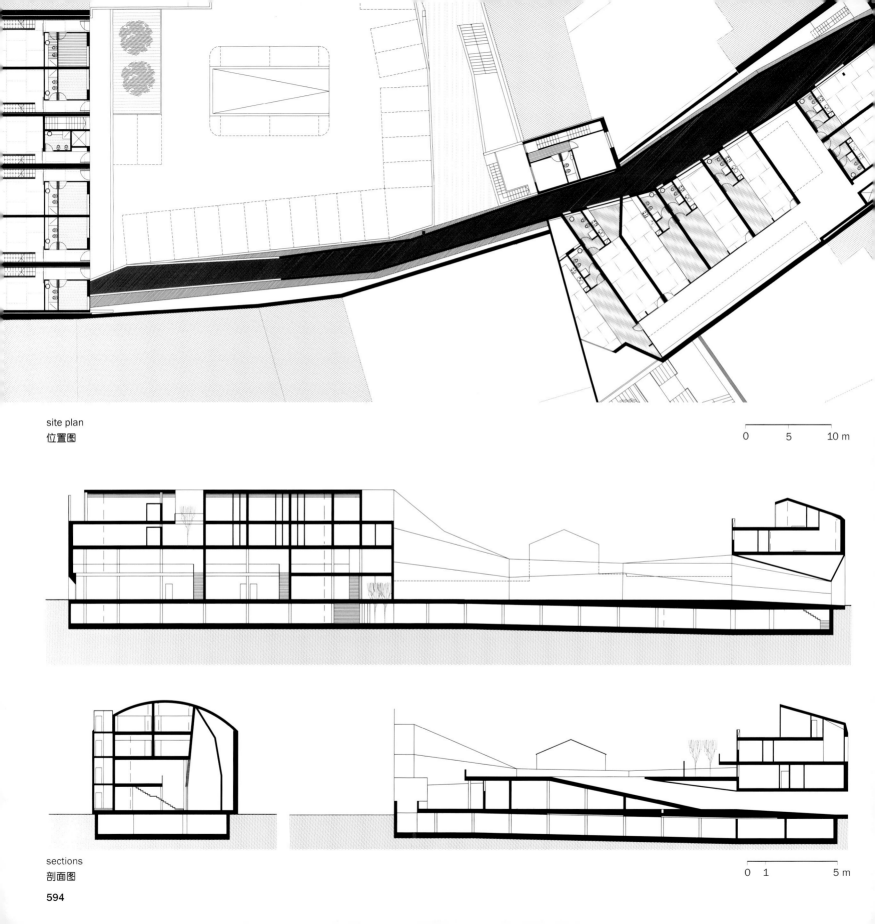

site plan
位置图

0   5   10 m

sections
剖面图

0   1   5 m

NURAGIC
ART MUSEUM

Cagliari, Italy, 2006

地点　卡利亚里，意大利
项目　博物馆
客户　Autonomous Region of Sardinia
结构　Studio Chessa
系统　Milano Progetti S.p.A.
规划　2006年设计竞赛，评审委员会推荐
成本　36,000,000欧元
建筑面积　12,050平方米

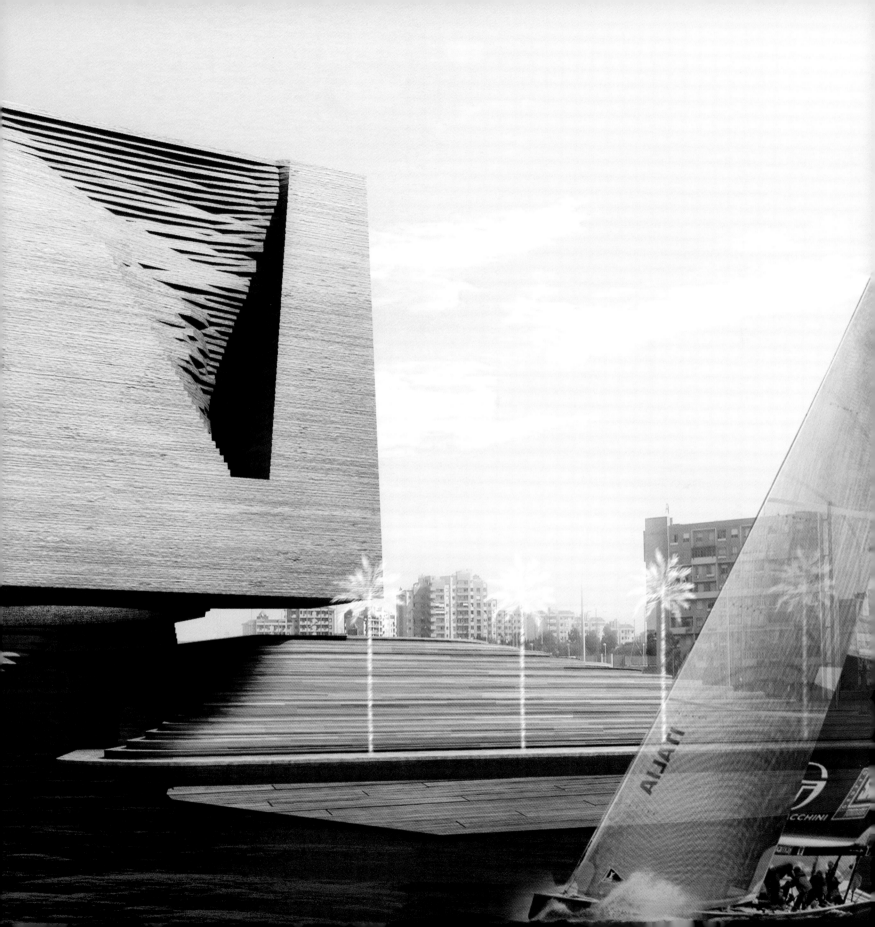

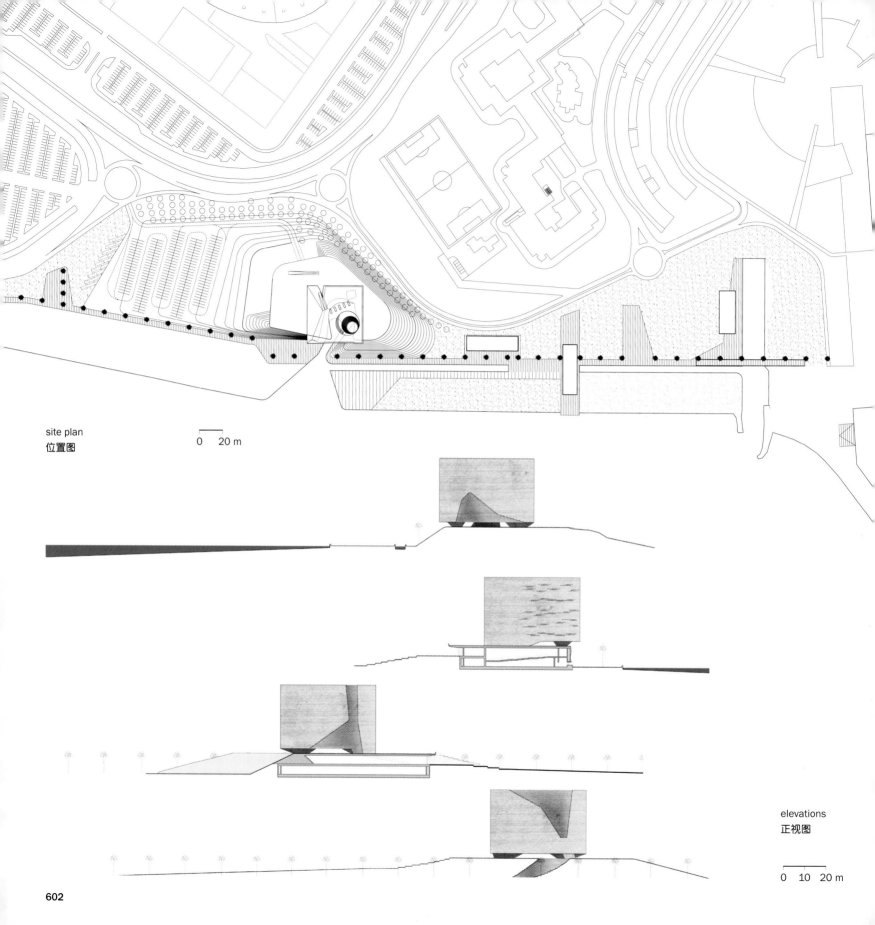

site plan
位置图

0 20 m

elevations
正视图

0 10 20 m

602

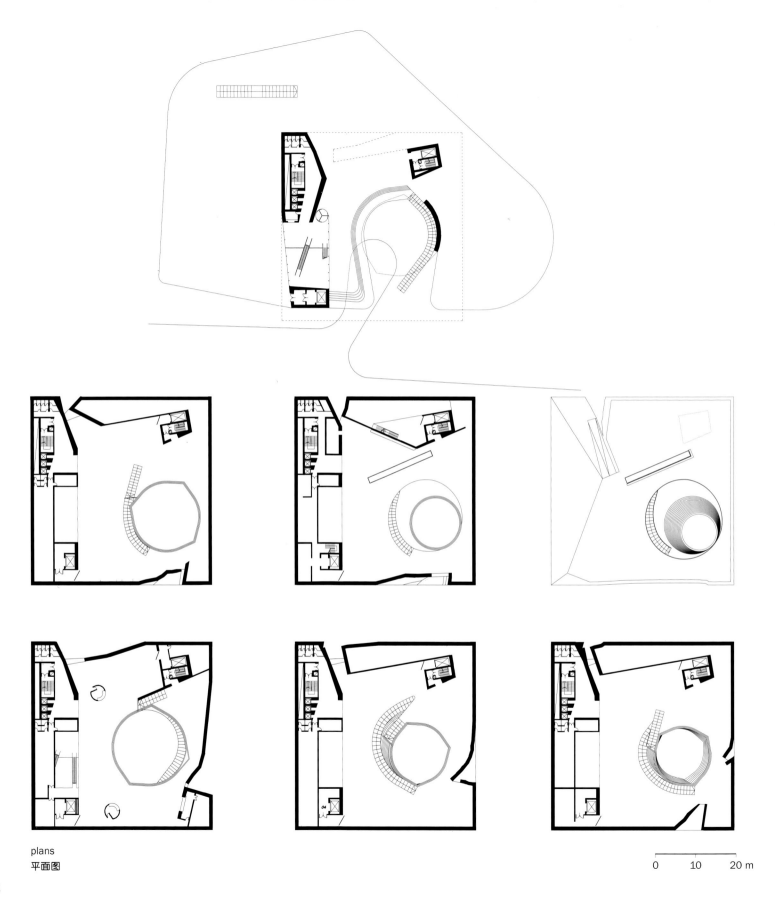

plans
平面图

0   10   20 m

606

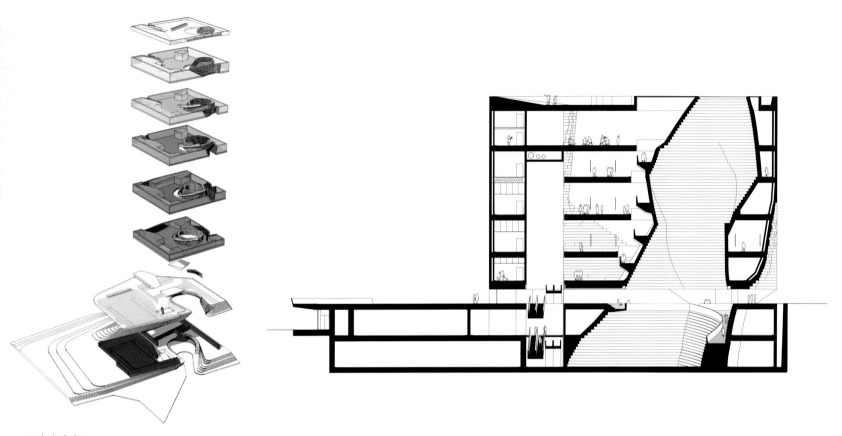

exploded view
分解图

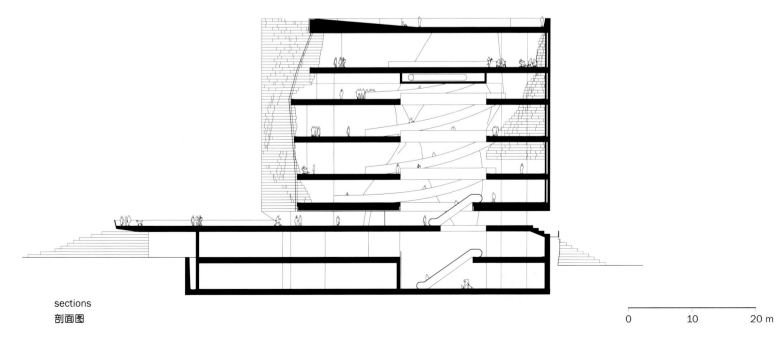

sections
剖面图

0   10   20 m

TORRE DELLE ARTI

residential and commercial building
Milan, Italy, 2006

地点　米兰，意大利
项目　住宅和商业楼
客户　Babcock & Brown
结构　Favero&Milan Ingegneria
系统　Studio Ti
规划　2006年-2007年
成本　50,000,000欧元
建筑面积　19,190平方米
体量　60,000立方米
承包商　Else costruzioni S.p.A. - Pessina Costruzioni, S.p.A.

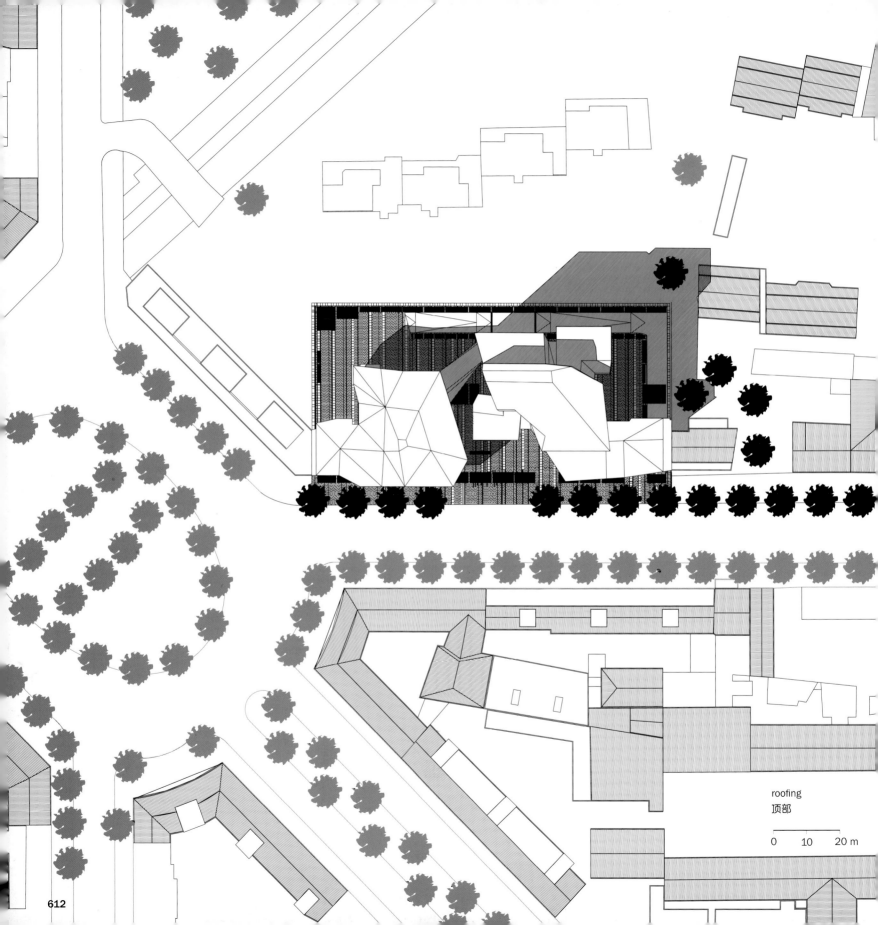

roofing
顶部

0   10   20 m

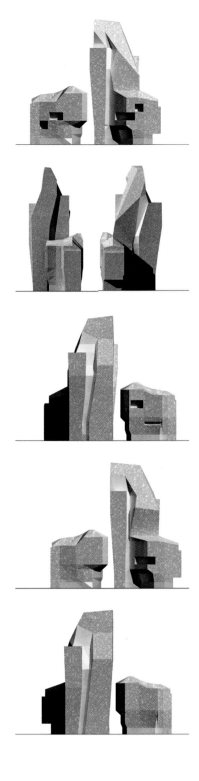

elevations development
正视图发展

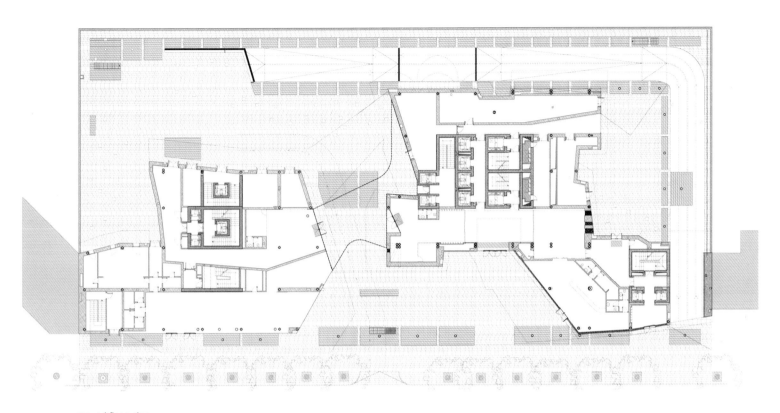

ground floor plan
一层平面

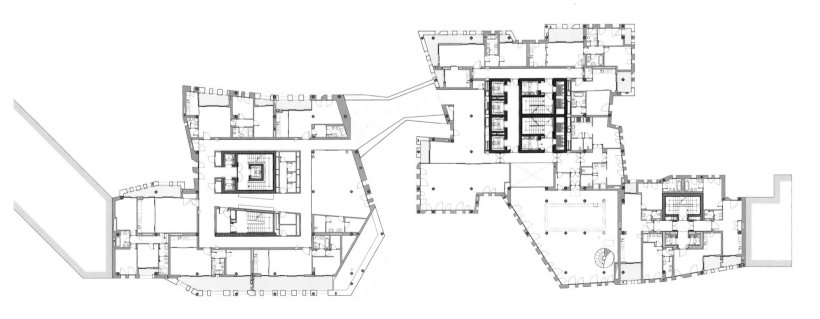

fourth floor plan
四层平面

0   5   10 m

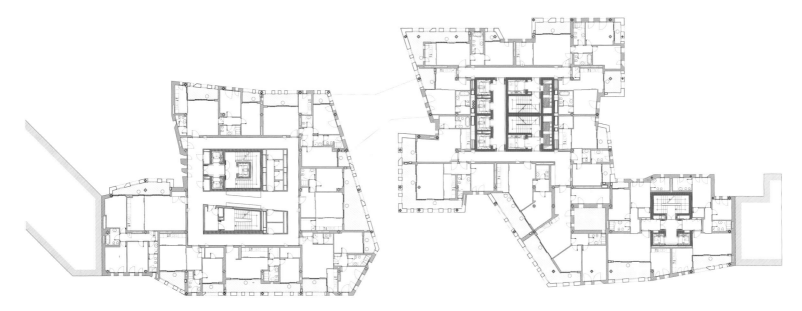

seventh floor plan
八层平面

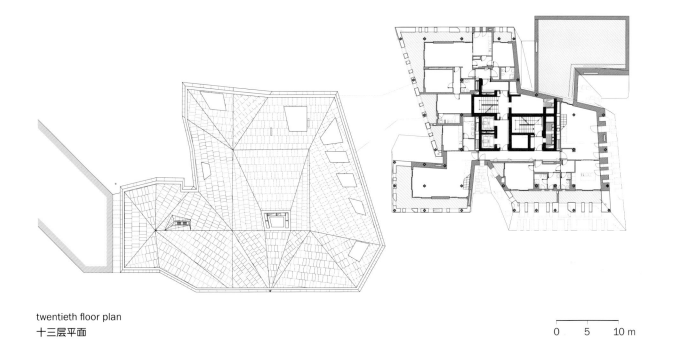

twentieth floor plan
十三层平面

0　　5　　10 m

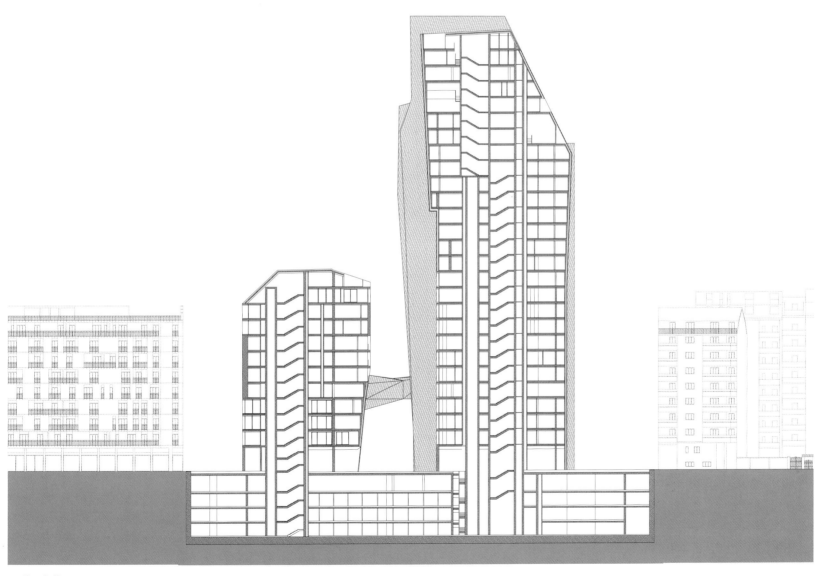

section A-A'
A-A'剖面

0   10   20 m

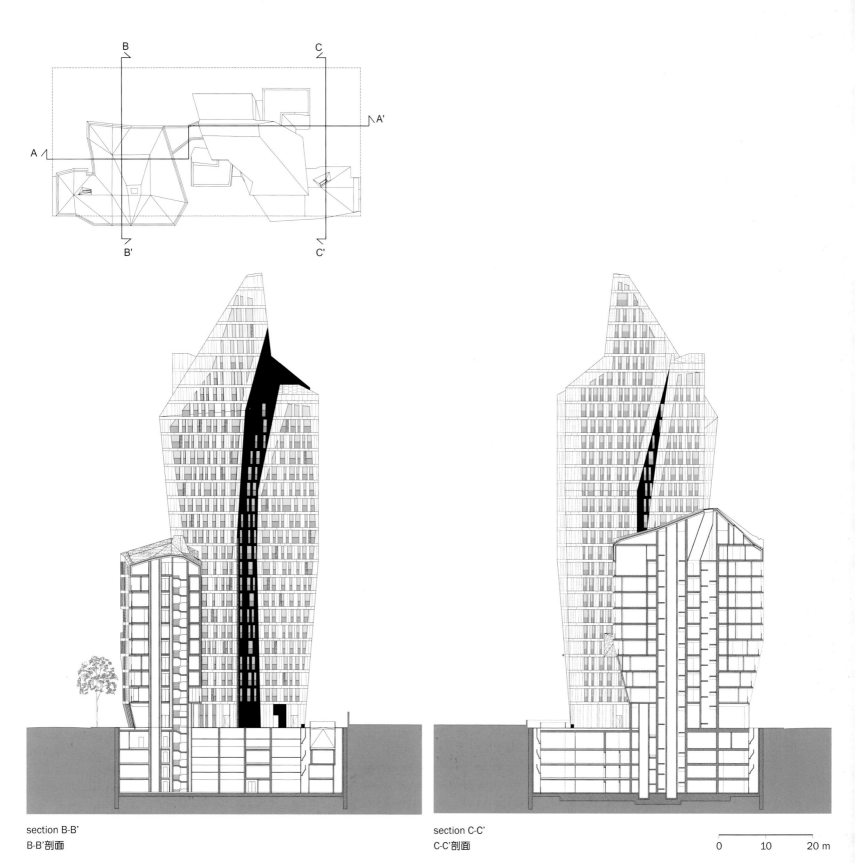

section B-B'
B-B'剖面

section C-C'
C-C'剖面

0   10   20 m

图书在版编目（CIP）数据

可持续性地标建筑 ：汉英对照 / 石大伟 主编. --北京 ： 中国林业出版社，2012.6

ISBN 978-7-5038-6603-6

Ⅰ．①可… Ⅱ．①石… Ⅲ．①建筑设计－作品集－世界－现代 Ⅳ．①TU206

中国版本图书馆CIP数据核字（2012）第094637号

可持续性地标建筑 （中）　　　　　　　　　　　　　　　石大伟　主编

责任编辑：李　顺
出版咨询：（010）83223051

出　版：中国林业出版社（100009 北京西城区德内大街刘海胡同7号）
印　刷：北京时捷印刷有限公司
发　行：新华书店北京发行所
电　话：（010）83224477
版　次：2012年6月第1版
印　次：2012年6月第1次
开　本：787mm×1092mm 1 / 12
印　张：76
字　数：200千字
定　价：880.00元（上、中、下册）